T0322082

Practical Optical Interferometry

Imaging at Visible and Infrared Wavelengths

Optical interferometry is a powerful technique to make images on angular scales hundreds of times smaller than is possible with the largest telescopes. This concise guide provides an introduction to the technique for graduate students and researchers who want to make interferometric observations, and acts as a reference for technologists building new instruments. Starting from the principles of interference, the author covers the core concepts of interferometry, showing how the effects of the Earth's atmosphere can be overcome using closure phase, and the complete process of making an observation, from planning to image reconstruction. This rigorous approach emphasises the use of rules-of-thumb for important parameters such as the signal-to-noise ratios, requirements for sampling the Fourier plane and predicting image quality. The handbook is supported by web resources, including the Python source code used to make many of the graphs, as well as an interferometry simulation framework, available at www.cambridge.org/9781107042179.

DAVID F. BUSCHER is a lecturer at the Cavendish Laboratory, University of Cambridge, and is System Architect for the Magdalena Ridge Observatory Interferometer, an imaging interferometer under construction in New Mexico. He works on the design, construction and exploitation of optical interferometers and adaptive optics systems and is the UK representative on the Science Council for the European Interferometry Initiative.

Cambridge Observing Handbooks for Research Astronomers

Today's professional astronomers must be able to adapt to use telescopes and interpret data at all wavelengths. This series is designed to provide them with a collection of concise, self-contained handbooks, which cover the basic principles peculiar to observing in a particular spectral region, or to using a special technique or type of instrument. The books can be used as an introduction to the subject and as a handy reference for use at the telescope or in the office.

Series Editors
Professor Richard Ellis, Astronomy Department, *California Institute of Technology*
Professor John Huchra, Center for Astrophysics, *Smithsonian Astrophysical Observatory*
Professor Steve Kahn, Department of Physics, *Columbia University*, New York
Professor George Rieke, Steward Observatory, *University of Arizona*, Tucson
Dr Peter B. Stetson, Herzberg Institute of Astrophysics, *Dominion Astrophysical Observatory*, Victoria, British Columbia

Practical Optical Interferometry
Imaging at Visible and Infrared Wavelengths

DAVID F. BUSCHER
University of Cambridge

CAMBRIDGE
UNIVERSITY PRESS

Shaftesbury Road, Cambridge CB2 8EA, United Kingdom

One Liberty Plaza, 20th Floor, New York, NY 10006, USA

477 Williamstown Road, Port Melbourne, VIC 3207, Australia

314–321, 3rd Floor, Plot 3, Splendor Forum, Jasola District Centre, New Delhi – 110025, India

103 Penang Road, #05–06/07, Visioncrest Commercial, Singapore 238467

Cambridge University Press is part of Cambridge University Press & Assessment,
a department of the University of Cambridge.

We share the University's mission to contribute to society through the pursuit of
education, learning and research at the highest international levels of excellence.

www.cambridge.org
Information on this title: www.cambridge.org/9781107042179

© Cambridge University Press & Assessment 2015

First published 2015

A catalogue record for this publication is available from the British Library

Library of Congress Cataloging-in-Publication data
Buscher, David F. (David Felix)
Practical optical interferometry imaging at visible and infrared wavelengths / David F.
Buscher, University of Cambridge.
pages   cm
Includes bibliographical references and index.
ISBN 978-1-107-04217-9
1. Optical measurements.   2. Interferometry.   3. Optical interferometers.
4. Astronomy.   I. Title.   II. Title: Optical interferometry imaging at visible and infrared
wavelengths.
QC367.3.I58B87   2015
522´.6–dc23
2015002196

ISBN   978-1-107-04217-9   Hardback

Additional resources for this publication at www.cambridge.org/9781107042179

Contents

Principal symbols, functions and operators

$\boldsymbol{B}_{ij}$	baseline vector between telescopes i and j
$F(\boldsymbol{u})$	coherent flux of object at spatial frequency $\boldsymbol{u}$
F_{ij}	coherent flux of fringes measured between telescopes i and j
F_i	flux measured through telescope i
$\mathcal{F}$	Fourier transform operator
$I(\sigma)$	object brightness at angular coordinate σ
$i(x)$	fringe intensity at coordinate x
i_p	fringe intensity at pixel p
$\text{jinc}(x)$	$J_1(x)/x$ where J_1 is the order-1 Bessel function of the first kind
P_{ij}	power spectrum of fringes measured between telescopes i and j
r_0	seeing coherence length (Fried parameter)
$\hat{\boldsymbol{S}}_0$	direction of the phase centre
SNR	signal-to-noise ratio
$\text{rect}(x)$	rectangular 'top-hat' function
s_{ij}	spatial frequency of fringes between telescopes i and j
T_{ijk}	triple product (bispectrum) of fringes measured on telescopes i, j and k
t_0	seeing coherence time
$\boldsymbol{u} = (u, v)$	projected baseline coordinate in wavelengths
$V(\boldsymbol{u})$	complex visibility of object at spatial frequency $\boldsymbol{u}$
V_{ij}	complex visibility of fringes measured between telescopes i and j
$\delta(\boldsymbol{x})$	Dirac delta function
η_i	complex gain coefficient for telescope i
γ_{ij}	complex visibility degradation for fringes measured between telescopes i and j
λ	optical wavelength

Λ_p	integrated classical intensity in pixel p
ν	optical frequency
Ψ	complex wave amplitude
σ	standard deviation
$\sigma = (l, m)$	angular coordinate with respect to phase centre

List of abbreviations

ALMA	Atacama Large Millimeter/submillimeter Array
AMBER	Astronomical Multi-BEam combineR
AO	adaptive optics
APD	Avalanche Photo-Diode
BSMEM	BiSpectrum Maximum Entropy Method
CCD	charge-coupled device
CHAMP	CHARA Michigan Phase tracker
CHARA	Center for High Angular Resolution Astronomy
COAST	Cambridge Optical Aperture Synthesis Array
DFT	discrete Fourier transform
ESO	European Southern Observatory
FFT	fast Fourier transform
FITS	Flexible image transport system
FLUOR	Fibered Linked Unit for Optical Recombination
FWHM	full width at half maximum
LBT	Large Binocular Telescope
MROI	Magdalena Ridge Observatory Interferometer
NPOI	Navy Precision Optical Interferometer
OIFITS	optical interferometry FITS
OPD	optical path difference
P2VM	pixel-to-visibility matrix
PIONIER	Precision Integrated-Optics Near-infrared Imaging ExpeRiment
PSF	point-spread function
PTI	Palomar Testbed Interferometer
resel	resolution element
RMS	root mean square
SNR	signal-to-noise ratio
SUSI	Sydney University Stellar Interferometer

SVD	singular value decomposition
V2PM	visibility-to-pixel matrix
VEGA	Visible spEctroGraph and polArimeter
VLA	Very Large Array
VLTI	Very Large Telescope Interferometer

Foreword

I was honoured and delighted when David Buscher invited me to write an introduction to his new book. I have long felt that there is a desperate need for an authoritative and accessible book on the techniques of optical and infrared interferometry and David has filled this major gap in the literature with this excellent piece of technical and scientific writing.

For the radio astronomer, interferometry is the bread and butter of how much of the discipline has to be undertaken. Radio, and nowadays millimetre and submillimetre, astronomers are brought up with the concepts of amplitude and phase, Fourier inversion and so on, which has always been something of a barrier to the wider appreciation of these disciplines by optical astronomers, who until recently scarcely had to bother about phase at all. The understanding of aperture synthesis imaging in all its variants became a black-belt sport for the initiates and this discouraged the typical astronomer from taking the plunge.

But this is no longer reasonable or acceptable. The possibilities opened up by optical and infrared synthesis imaging are enormous, as David makes clear in this book. Angular resolution of a milliarcsecond or better can now be routinely provided by the most advanced optical-infrared interferometers and will undoubtedly result in important new discoveries and much improved tests of theories of Galactic and extragalactic objects.

This is where David's book comes in. He offers a rigorous, but accessible, introduction to the necessary theoretical and experimental tools needed to understand and apply the techniques of optical synthesis imaging. The result is that the effort needed to understand the key concepts by those new to the field, or who are still put off by the apparent complexity of the techniques, is made very much less forbidding. There is so much remarkable astrophysics lurking below the 1 milliarcsecond threshold that it is only a question of time before optical-infrared interferometry becomes a standard tool of the trade. David's excellent exposition makes this feasible for all astronomers and I warmly recommend this book to all of them.

Malcolm Longair
October, 2014

Preface

Optical interferometry uses the combination of light from multiple telescopes to allow imaging on angular scales much smaller than is possible with conventional single-telescope techniques. It is increasingly recognised as the only technique capable of answering some of the most fundamental scientific questions in astronomy, from the origin of planets to the nature of the physical environments of black holes.

Interferometry is an established technique at radio and millimetre wavelengths, with instruments such as the VLA and ALMA being the workhorses at these wavelengths. The development of interferometry in the optical (which we take here to include both visible and infrared wavelengths) has lagged behind that of radio interferometry due both to the extreme precision requirements imposed by the shorter wavelengths and to the severe effects of the Earth's atmosphere. For many years, the use of optical interferometry for scientific measurements was limited to the specialists who designed and built interferometric instruments.

At the beginning of the twenty-first century, the first "facility" optical interferometers such as the VLTI, the CHARA array the Keck Interferometer came online, with the aim of broadening the use of interferometry to the wider community of astronomers who could use it as a tool to do their science. As part of this expansion, organisations in Europe and the USA began to hold summer schools to provide an introduction to the theory and practice of interferometry to astronomers new to the topic. A number of times after giving lectures at these schools, I have had students come up to me wanting to find out more about some 'well-known' interferometric idea that I have mentioned in my talk. Often I have had to reply that there is no one place in the literature which provides this further information.

While a lot of the knowledge has been built up by the community involved in the development of optical interferometry, much of it is not written down, or is scattered over multiple separate papers. The most useful single reference on the topic is provided by the course notes from the 1999 Michelson Summer School (Lawson, 1999), but of necessity it cannot cover the decade and more of developments in interferometry since its publication. Knowledge relevant to optical interferometry is also available in radio-astronomy textbooks (a good example is *Interferometry and Synthesis in Radio Astronomy* (Thompson *et al.*, 2008)), but finding it requires the reader to have an existing understanding of which elements of radio interferometry are relevant in practice to optical interferometry. This is because, although the same physical principles apply at both radio and optical wavelengths, many concerns which are important to the practice of interferometry in one wavelength regime do not arise in the other.

This book is mainly aimed at enabling the newcomer to optical interferometry to get 'up to speed' on the topic, providing a basic reference on the fundamental concepts applicable to present-day optical interferometry. It aims to give simple rules-of-thumb to allow a quick understanding of what is important, but also shows how a more rigorous understanding of more detailed aspects can be derived, without necessarily giving detailed derivations in all cases. The ideas are presented assuming a mathematical background at the level required for an undergraduate physics course, including concepts such as complex exponentials and random walks. Where possible, examples from the literature are given in order to relate the more abstract ideas to their practical roots.

The main intended audience for this book is students and researchers in astronomy who want to use an interferometer as a tool for doing science. This book concentrates on the most 'mainstream' application of interferometry, which is loosely termed 'imaging': this can range from measuring a few parameters of an object's appearance based on a simple model (for example measuring a diameter of a star under the assumption that it is round) to true 'aperture synthesis', reconstructing model-independent images of the object under study.

In the interests of brevity, less is said here about interferometric techniques such as nulling, polarimetry and astrometry that have undoubted scientific potential, but are less mature in their application, and as a result less readily available as a science tool for non-expert users. These techniques are evolving rapidly, and the consensus as to the best way to make them work is still quite fluid. Any useful discussion of these topics would need to cover all the possible directions of development in these modes of interferometry and might still not cover the key ideas which will turn out to be of importance to their successful establishment.

A second and overlapping audience for this book is students, instrumentation scientists and engineers who want to work on developing and building new instrumentation for interferometry. Interferometry is clearly on an upward arc, with new initiatives such as the Planet Formation Imager (http://www.planetformationimager.org) gaining support from a wide scientific user base. These initiatives will require the efforts of a new generation of scientists and technologists who are well-versed in the existing ideas of interferometry, but are also able to see beyond these techniques to understand what can make interferometry even more useful in the future. This book will hopefully provide a small stepping stone for such people.

Acknowledgements

I would like to thank my wife for her encouragement to take the idea for this book forward and her support for and patience with the process of writing it. The majority of the book's diagrams were drawn by Fran Harwood-Whitcher, who turned my rough scrawls into models of clarity. Karen Scrivener was instrumental in getting the necessary permissions for the images borrowed from other publications. James Gordon, Malcolm Longair, Aglaé Kelerer and Alex Wells read drafts of the manuscript and provided helpful comments and discussion. I am grateful to my colleagues at the Cavendish Laboratory for providing a stimulating work environment and the sabbatical which allowed me to get this book started. My editor, Vince Higgs, gave helpful advice along the way, and my colleagues in the worldwide interferometry community provided me with gentle and useful reminders that I really ought to finish this book. I would like to thank my mother for instilling me with an appreciation of the rewards of learning and teaching, and my father for his infectious enthusiasm for ideas and their practical application; the combination of these values in many ways provided the origin for this book.

1

Making fringes

1.1 The need for angular resolution

Progress in astronomy is dependent on the development of new instrumentation that can provide data which are better in some way than the data which
were available before. One improvement that has consistently led to astronomical discoveries is that of seeing finer detail in objects. In astronomy, the
majority of the objects under study cannot easily be brought closer for inspec-
tion, so the typical *angular* scales subtended by objects, which depend on the
ratio of the typical sizes of the objects to their typical distances from the Earth,
are a more useful indicator of how easily they can be seen than their linear
sizes alone. The angular separation of two features in a scene which can be
just be distinguished from one another is called the *angular resolution* and the
smaller this scale is, the more detail can be seen.

The impact of increased angular resolution can be appreciated from comparing the important angular scales of objects of interest with the angular
resolution of the instrumentation available at different times in history. Prior
to the invention of the telescope, the human eye was the premier 'instrument' in astronomy, with an angular resolution of about 1 arcminute (about
300 microradians). With the notable exceptions of the Sun, Moon and comets,
most objects visible in the night sky are 'star-like': they have angular sizes
smaller than 1 arcminute and so appear as point-like objects. The first telescopes improved the angular resolution of the naked eye by factors of three to
six: Galileo's telescopes are thought to have had angular resolutions of about
10–20 arcseconds (Greco *et al.*, 1993; Strano, 2009) and it became possible
to see that planets appear as discs or crescents and have their own moons.
Subsequent improvements to telescopes have culminated in telescopes like
the Hubble Space Telescope (HST) which have typical angular resolutions of
around 50 milliarcseconds (about 250 nanoradians) – better by a factor of more
than a thousand than the naked eye.

1

The increased angular resolution offered by the HST and other high-angular-resolution telescopes has transformed the study of astrophysics. Astronomers have seen many phenomena that were undreamed of even a few decades earlier: young stars surrounded by discs of material left over from their formation, hugely complex filamentary structure in the 'planetary nebulae' surrounding stars at the end of their lives and bright 'cusps' of stellar emission at the centres of galaxies indicating the presence of black holes.

Nevertheless, the angular resolution available with a conventional optical telescope is still inadequate to resolve many important astrophysical phenomena. Amongst the most obvious examples are the following:

Stars – The photospheres of the nearest stars (except for the Sun) are a few milliarcseconds across.

Planet formation – A planet in an Earth-like orbit forming around a star in the nearest star-forming region (around 150 parsecs away) will be about 6 milliarcseconds from its parent star.

Black-hole accretion – The standard model for active galactic nuclei consists of an accreting black hole surrounded by a broad-line region which reprocesses the radiaton emerging from the accretion disc, and a torus of dusty material which can block direct radiation from the accretion disk. The dust tori in the nearest active galactic nuclei have angular diameters of a few milliarcseconds and the broad-line regions are predicted to have sub-milliarcsecond angular radii. The accretion disks themselves are thought to have *micro* arcsecond-scale diameters.

Undeniably, then, there is scope for observing new phenomena if angular resolutions much greater than those available with current telescopes could be achieved.

1.2 The resolution of a single telescope

If a telescope is built so that all optical imperfections are overcome and the distorting effects of the Earth's atmosphere are removed (for example by placing the telescope in space), then the angular resolution of the telescope will be limited by diffraction. This can be understood by considering the observation of a point source of light using such an idealised telescope.

The telescope can be modelled as a perfect lens projecting an image onto a detector as shown in Figure 1.1. The finite size of the telescope is modelled as a circular aperture of diameter d placed in front of the lens. When observing

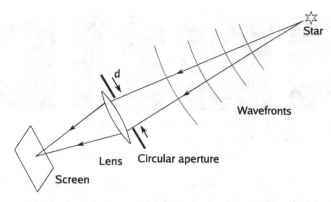

Figure 1.1 A telescope focussing the light from a point source of light (e. g. a star) onto a screen. The telescope is represented as a perfect lens with a circular aperture of diameter d in front of it.

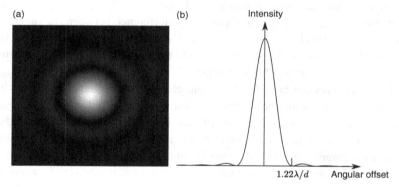

Figure 1.2 The diffraction-limited intensity pattern (known as the 'Airy disc') seen on the screen in the focal plane of the telescope in Figure 1.1 (a) and a cut through the intensity pattern (b).

a point-like object (which will be referred to as a 'star', since most stars are close enough to point-like for these purposes), this arrangement corresponds to a Fraunhofer diffraction experiment. What is seen on the screen is not an infinitely sharp point of light but rather the diffraction pattern of the circular aperture, known as an *Airy disc*, which consists of a central spot surrounded by circular rings as shown in Figure 1.2. This diffraction pattern is known as the *point-spread function* (PSF) of the telescope. Diffraction therefore introduces a finite amount of 'blurring' to the image of the point-like source, even though the lens is modelled as being free from any defects.

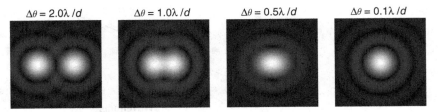

Figure 1.3 Patterns seen in the focal plane of a telescope when pairs of stars of different separations $\Delta\theta$ are observed through a telescope.

The effect of the blurring on angular resolution can be quantified by considering what happens if there is a second star close to the first one as shown in Figure 1.3. The light from the second star will produce a second identically shaped diffraction pattern on the screen, offset by an angular distance $\Delta\theta$ where $\Delta\theta$ is the angular separation of the two stars. Since the phase of the light waves from one star varies randomly and independently of the phase of the light waves from the other star, there is no interference between the two (the justification for this lack of interference is discussed further in Section 1.4.3), and so what is seen on the screen is simply the sum of the intensity patterns that would be seen with either star alone.

If the two stars are brought closer and closer to one another as shown in Figure 1.3, then at some point it becomes impossible to tell whether there is one star or two. At this point the pair of stars is said to be 'unresolved' and the separation at which this occurs is the angular resolution of the telescope. While the exact separation at which the stars become indistinguishable depends on a number of factors such as their relative brightnesses, the *Rayleigh criterion* defines the stars as being 'just resolved' when the peak of the blur pattern produced by one star overlaps with the first null of the blur pattern produced by the second. As shown in Figure 1.2, the angular distance from the peak to the first null of the Airy disc is given by $1.22\lambda/d$, where λ is the wavelength of the light being observed. Thus, the required overlap occurs when

$$\Delta\theta = 1.22\lambda/d. \tag{1.1}$$

By the Rayleigh criterion, Equation (1.1) gives value of the angular resolution of any sufficiently well-corrected telescope. As an example, we can consider the HST, which has a 2.4-m-diameter primary mirror. When observing at a visible wavelength of 500 nm, the diffraction spot from this telescope will have an angular radius of 52 milliarcseconds and so two stars closer together than this cannot be reliably distinguished.

The angular resolution can be improved by building larger telescopes. However, to achieve 1 milliarcsecond resolution at a wavelength of 500 nm would require a telescope 126 m in diameter. Even the largest optical telescopes being proposed at present will have aperture diameters of less than 40 metres, and they come with billion-dollar price tags. The cost of a telescope scales as the square or cube of the aperture diameter so building telescopes more than three times as large seems unlikely in the medium term.

What is required is a method of gaining large factors of improvement in angular resolution which do not require unfeasibly large telescopes. The only known method with this property is long-baseline interferometry. Interferometry uses the interference of light from two or more small telescopes separated by a large distance B to get images with an angular resolution of order λ/B without the mechanical and optical complexities inherent in constructing a single large telescope of diameter B. The next sections serve to show the principles of this method.

1.3 A long-baseline interferometer

An astronomical interferometer collects the light originating from a single region of the sky at two or more locations and brings the collected beams of light together to form an interference pattern. Figure 1.4 shows perhaps the simplest interferometer which could be implemented in practice. Most real interferometers include magnifying and/or demagnifying optics to make the construction of the interferometer easier and cheaper, but the design shown has the advantage that it achieves all the essential functions of an interferometer using only plane (flat) mirrors and so the optical functions of all the elements are readily understandable.

The example interferometer collects starlight using a pair of siderostats, flat mirrors which can be tilted appropriately to reflect the light from a chosen region of sky into a fixed direction. The diameter of the siderostat mirrors can be modest, perhaps only 5 cm if only bright objects are to be observed. In an interferometer used for studying faint objects, the siderostats would typically be replaced by individual telescopes acting as light collectors, each perhaps several metres in diameter.

The distance between the two collectors is typically much larger than the size of any feasible individual collector, perhaps hundreds of metres. The orientation in space of the collector separation is also important: the *baseline vector* $\boldsymbol{B}_{pq}$ between the light-collecting elements p and q of an interferometer is defined as $\boldsymbol{B}_{pq} = \boldsymbol{x}_p - \boldsymbol{x}_q$, where the elements are situated at locations $\boldsymbol{x}_p$

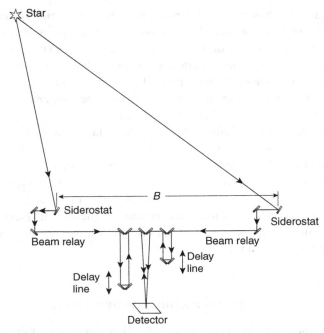

Figure 1.4 A simple long-baseline interferometer constructed out of plane (flat) mirrors. Some of the mirrors have been included to maintain certain symmetries of the optical path – these symmetries are explained in Section 4.4.2.

and x_q. In much of the following the pq suffix is dropped, but it will be used later when considering multi-baseline interferometers. The length of the baseline vector is called the 'baseline length' or just the 'baseline' and is shown as B in Figure 1.4.

The collected starlight is brought to a central point using reflections off a series of 'beam-relay' mirrors. Included in the beam path are a pair of mirrors, which can be moved backwards or forwards, acting like an 'optical trombone' to vary the distance the light travels before reaching the central combination point. These 'path compensators' or 'delay lines' serve to control the relative delays between the light beams coming from different collectors: as will be discussed in Section 1.7, in practice it is necessary for the times taken for the light to travel from the object to the point of interference via the two collectors to be matched with one another to in order to see interference.

The light beams are combined in a 'beam combiner' to produce interference fringes. There are many arrangements which can be used to do this: the arrangement shown here uses a so-called 'pupil-plane' arrangement where the two beams are simply allowed to overlap on a screen. The intensity pattern on the screen can be observed visually but it is more usual to replace the screen by an electronic detector in order to obtain more quantiative information and to observe fainter objects. The detector converts the intensity at each location on its face into an electronic signal, which is digitised, analysed and displayed on a computer.

Interference between the two beams results in a sinusoidal intensity pattern. In the next section it will be demonstrated that this 'fringe pattern' contains information about the size and shape of the object being observed.

1.4 The interferometric measurement equation

The properties of the fringe pattern seen when observing complex objects is derived in the following analysis by considering first the fringe pattern formed when observing a point source, and then how the characteristics of the fringe pattern change when the object consists of two closely spaced point sources. Finally, the properties of the fringe pattern formed when observing an arbitrary object will then be derived by considering it as a collection of closely spaced point sources.

1.4.1 The fringe pattern from a point source

The form of the fringe pattern is derived here using a model of the interferometer which is shown schematically in Figure 1.5. In this model, light arrives from a source of light that is the object of interest. The source is assumed to be a point-like 'star', which is sufficiently distant that the light can be accurately represented as a plane wave, in other words there are plane surfaces known as 'wavefronts' over which the instantaneous electromanetic field E_0 is the same at any given moment in time.

Light propagates from this wavefront along two parallel rays and arrives at the two collectors. The rays then travel via the interferometer optics to the beam combination point, at which point the light waves are superposed and then converted into an intensity $i(x)$. These rays are subject to a series of delays consisting of three different components:

1. An 'external' or 'geometric' delay τ_{ext} due to the light travel time from the wavefront to the collector.
2. An 'internal' delay τ_{int} due to the light travel time along the beam-relay and delay-line beam paths.

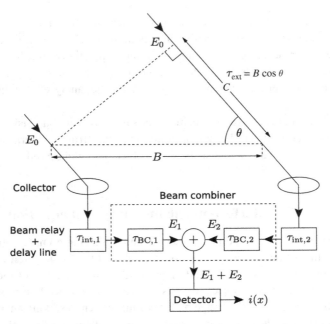

Figure 1.5 A simplified model of fringe formation in an interferometer.

3. A beam-combiner delay $\tau_{BC}(x)$, which is dependent on the location x on the detector on which the beam lands, as shown in Figure 1.6. It will be seen in the following analysis that an important function of any beam-combiner design is to allow the sampling of the interference patterns at multiple locations in 'delay space' and in this case these locations are dependent on the detector coordinate x.

This model neglects the effects of light losses in the interferometer. It also neglects other effects such as optical imperfections along the beam path because it assumes that all rays passing through one collector and arriving at the entrance of the beam combiner experience the same delay. The benefits of using this model are that the analysis is simpler and, perhaps more importantly, the results can be readily applied to interferometers of different designs. For example, the model can be straightforwardly applied to an interferometer which uses temporal coding of the fringes (see Section 4.7) instead of spatial fringes by replacing the detector coordinate x with a time coordinate t.

The analysis starts by considering the electromagnetic field incident on the interferometer. The light is assumed to be perfectly monochromatic so the light wave consists of an electromagnetic wave oscillating at frequency v and the

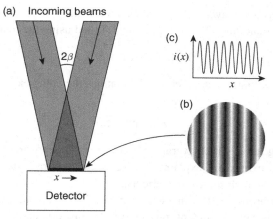

Figure 1.6 (a) Beams of starlight arriving at angles of $\pm\beta$ on a detector. (b) The fringe pattern on the detector. (c) A one-dimensional cut through the fringe pattern.

therefore instantaneous electric field at any point on the initial wavefront is given by

$$E_0 = \frac{2}{\epsilon_0}\mathrm{Re}\left[\Psi_0 e^{-2\pi i\nu t}\right], \tag{1.2}$$

where ϵ_0 is the electric permittivity of free space and Ψ_0 is a 'complex wave amplitude' given by

$$\Psi_0 = |\Psi_0|\,e^{i\phi_0}, \tag{1.3}$$

where ϕ_0 is the phase of the wave. The electric field is not represented as a vector in this 'scalar wave' analysis. It is assumed that the properties of the system being analysed are the same for any polarisation, and so the vector properties of the electromagnetic field are ignored.

At optical frequencies, the oscillations of the wave are typically not directly observable: optical detectors effectively measure the accumulated light energy received over an 'exposure time' or 'integration time', which can be anywhere from several picoseconds to many minutes, whereas the oscillation period is a few femtoseconds. What is observable is the mean intensity of the wave (i. e. the mean energy crossing unit area per unit time, also known as the 'flux' from the object) given by

$$F_0 = \left\langle \epsilon_0 E_0^2 \right\rangle, \tag{1.4}$$

where $\langle\rangle$ represents averaging over the integration time of the detector. Substituting Equation (1.2) into Equation (1.4) and using the relationship

$$\text{Re}\,\{X\} = \tfrac{1}{2}(X + X^*), \tag{1.5}$$

where X is any complex number and X^* denotes the complex conjugate of X, gives

$$F_0 = |\Psi_0|^2, \tag{1.6}$$

after dropping terms which average to zero over the exposure time. The simplicity of this expression explains the seemingly arbitrary factor of $2/\epsilon_0$ in Equation (1.2), which serves to define Ψ_0.

In an interferometer, the incident intensity F_0 is not measured directly. Instead, the wave is incident on the two collectors and travels via the beam-relay optics to the beam combiner where the beams are combined and arrive at a given location on the detector surface denoted by a coordinate x. At this location the electric field is given by the superposition of the two fields $E_1(x)$ and $E_2(x)$ that would have been observed from each collector alone. The mean intensity of the light received by the detector at a given location x is therefore given by

$$i(x) = \epsilon_0 \left\langle (E_1(x) + E_2(x))^2 \right\rangle$$
$$= \left\langle \left(\text{Re}\left[\Psi_1(x)e^{-2\pi i\nu t} + \Psi_2(x)e^{-2\pi i\nu t} \right] \right)^2 \right\rangle, \tag{1.7}$$

where $\Psi_1(x)$ and $\Psi_2(x)$ are defined analagously to Ψ_0.

Expanding and dropping terms which average to zero over the integration time of the detector gives

$$i(x) = |\Psi_1(x)|^2 + |\Psi_2(x)|^2 + 2\text{Re}\left[\Psi_1(x)\Psi_2^*(x) \right]. \tag{1.8}$$

The intensity is therefore the sum of the intensities which would be observed on either beam alone, plus a cross term which depends on a product of the two wave amplitudes. This 'interference term' can be positive or negative, corresponding to constructive and destructive interference respectively.

Assuming that the interferometer optics introduce no losses in the light intensity, then the waves arriving at the detector are simply time-delayed versions of the incident wave $E_0(t)$, given by

$$E_1(t) = E_0(t - \tau_{\text{ext},1} - \tau_{\text{int},1} - \tau_{\text{BC},1}(x)) \tag{1.9}$$

and

$$E_2(t) = E_0(t - \tau_{\text{ext},2} - \tau_{\text{int},2} - \tau_{\text{BC},2}(x)). \tag{1.10}$$

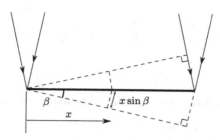

Figure 1.7 Geometry of the relative path lengths travelled by the beams arriving at the detector surface in Figure 1.6 as a function of the coordinate x.

The definition of the oscillating wave amplitude given in Equation (1.2) shows that the time delay between E_0 and E_1 and E_2 corresponds to phase shifts between the incoming wave amplitude Ψ_0 and the delayed wave amplitudes Ψ_1 and Ψ_2, so that

$$\Psi_1 = \Psi_0 e^{2\pi i \nu [\tau_{ext,1} + \tau_{int,1} + \tau_{BC,1}(x)]} \tag{1.11}$$

and

$$\Psi_2 = \Psi_0 e^{2\pi i \nu [\tau_{ext,2} + \tau_{int,2} + \tau_{BC,2}(x)]}. \tag{1.12}$$

Substituting Equations (1.11) and (1.12) into Equation (1.8) gives

$$i(x) = |\Psi_0|^2 + |\Psi_0|^2 + 2|\Psi_0|^2 \mathrm{Re}\left[e^{2\pi i \nu [\tau_{12} + \tau_{BC,1}(x) - \tau_{BC,2}(x)]}\right]$$

$$= 2F_0 \left(1 + \mathrm{Re}\left[e^{2\pi i \nu [\tau_{12} + \tau_{BC,1}(x) - \tau_{BC,2}(x)]}\right]\right) \tag{1.13}$$

where τ_{12} is the component of the delay difference, which is independent of the location on the detector:

$$\tau_{12} = (\tau_{ext,1} - \tau_{ext,2}) + (\tau_{int,1} - \tau_{int,2}). \tag{1.14}$$

It should be noted that the usual convention in optics is to express time-delay differences such as τ_{12} in terms of the equivalent *optical path difference* (OPD), which is the time-delay difference multiplied by the speed of light in a vacuum c. An OPD of 1 m corresponds to approximately 3.3 nanoseconds (ns) of delay difference and an OPD of 1 micron (μm) corresponds to 3.3 femtoseconds (fs). (For those who prefer to use Imperial units, 1 ns of delay corresponds quite closely to 1 ft of OPD.) In this book OPD and delay will both be used, depending on which is most convenient: in almost all cases when numerical values are quoted they will be in microns of OPD.

For the pupil-plane beam combiner given in the example interferometer, the beams arrive at the detector with angles of incidence of $\pm\beta$ at the detector, as shown in Figure 1.6(a). Figure 1.7 shows that the rays hitting the detector surface at location x travel an extra distance $\pm x \sin\beta$ compared to the rays hitting at location $x = 0$. The OPD between the two beams therefore varies with x as

$$c\left(\tau_{\mathrm{BC},1}(x) - \tau_{\mathrm{BC},2}(x)\right) = 2x \sin\beta. \qquad (1.15)$$

The intensity on the detector is therefore a sinusoidal pattern given by

$$i(x) = 2F_0\left(1 + \mathrm{Re}\left[e^{\mathrm{i}(2\pi s x + \phi_{12})}\right]\right), \qquad (1.16)$$

where s is the 'fringe frequency' or the 'spatial frequency of the fringes' and is given by

$$s = 2\nu \sin\beta/c = 2\sin\beta/\lambda, \qquad (1.17)$$

where $\lambda = c/\nu$ is the wavelength of the radiation, and ϕ_{12} is a phase offset given by

$$\phi_{12} = 2\pi\nu\tau_{12}. \qquad (1.18)$$

This pattern is shown in Figure 1.6 and is known as a 'fringe pattern'. The alternating dark and light stripes are called 'fringes' and are the characteristic sign of interference – when light from one of the collectors is blocked off, the 'stripes' will disappear leaving a uniform illuminated disc.

The peak-to-peak spacing of the fringe pattern is given by $1/s = \lambda/(2\sin\beta)$. Since λ is on the order of a micron, quite narrow angles of incidence β are needed in order to yield macroscopic-sized fringes: for $\lambda = 0.5\,\mu\mathrm{m}$, then $\beta = 50$ arcseconds will give a spacing between successive dark fringes of approximately 1 mm. This means that the distance between the beam-combining mirrors and the detector needs to be of order 100 m or more for 5-cm-diameter beams. In practice, different architectures of beam combiner, for example those employing beamsplitters or beam-reducing optics are used in order to allow the use of sensible-sized optics and optical paths (see Section 4.7).

1.4.2 Astrometric phase

The phase shift ϕ_{12} of the fringes depends on the external delay difference due to the difference in the optical paths from the star to the two collectors and the internal delay difference due to the difference in the optical paths from the collectors to the beam combiner. As shown in Figure 1.5, light from a star at infinity travels an additional distance external to the interferometer to get to one collector compared to the other. The delay difference due to this additional light path is given by

$$\tau_{\mathrm{ext},12} = \tau_{\mathrm{ext},1} - \tau_{\mathrm{ext},2} = B\cos\theta/c \qquad (1.19)$$

where B is the baseline and θ is the angle between the direction to the star and the baseline vector.

The delay lines can be used to adjust the internal delays so that the net delay τ_{12} and hence the fringe phase shift ϕ_{12} is zero for a star in a particular direction θ_0. If the delay is now kept fixed and a star offset from this 'phase centre' is observed, the OPD will be

$$\tau_{12}c = B\cos(\theta_0 + \Delta\theta) - B\cos\theta_0$$
$$\approx -\Delta\theta B \sin\theta_0, \tag{1.20}$$

where $\Delta\theta$ is the (small) angular offset of the star from the phase centre. The phase shift of the fringes will therefore be given by

$$\phi_{12} = -2\pi u\Delta\theta, \tag{1.21}$$

where u is the length of the projection of the baseline in the direction perpendicular to the star, scaled in units of wavelengths, i.e.,

$$u = B\sin\theta_0/\lambda. \tag{1.22}$$

Importantly, then, we can see that *the phase of the fringes is sensitive to the angular position of the source and this angular sensitivity increases with the length of the projected baseline.*

To get a numerical idea of this sensitivity, we can consider an interferometer operating at a wavelength of $\lambda = 500$ nm and a collector spacing of $B = 50$ m. For a phase centre which is nearly overhead so that $\sin\theta_0 \approx 1$ if the star position is shifted by $\Delta\theta \approx 1$ milliarcsecond (about 5 nanoradians) the phase of the fringes will shift by $180°$, so that where there was a bright fringe there is now a dark fringe.

An interferometer can therefore act as an 'angle meter', turning small changes in position of an object into easily visible shifts in the fringe pattern. At radio wavelengths, interferometry is the premier means of precise measurement of angular positions of celestial objects (known as astrometry), achieving sub-milliarcsecond precision (Ma *et al.*, 1998). A number of instruments using interferometry for astrometry at optical wavelengths have been proposed and built (Hummel *et al.*, 1994; Armstrong *et al.*, 1998; Colavita *et al.*, 1999; Shao, 1998; Launhardt *et al.*, 2007) but for a number of instrumental reasons single-telescope methods are still predominant in astrometry.

1.4.3 The fringe pattern from two point sources

Figure 1.8 shows the geometry for the light arriving at an interferometer from two distant point-like objects or 'stars'. One star has an incident electric field

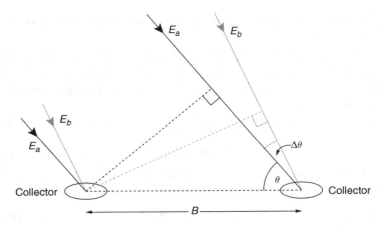

Figure 1.8　Light arriving from two stars with separation $\Delta\theta$.

E_a and is at the phase centre of the interferometer and the second star has an electric field of E_b and is at a small angular offset of $\Delta\theta$.

If $\Delta\theta$ is sufficiently small, four beams of light will arrive at the detector, one from each star via each collector. To a good approximation all the beams will fall directly on top of one another. In this case, six sets of interference effects need to be considered, as there are six possible pairwise combinations of four beams.

Superposition of intensities

In reality only the interference between pairs of beams originating from the same star will be seen. The reason for this is that the light emitted from a natural object is not a perfectly stable sinusoid as has been assumed up to this point. It is necessary instead to consider what is called a *quasi-monochromatic* light beam, where the wave amplitude Ψ is not constant but is varying randomly on a timescale which is long compared to any differential delays in the system, but short compared to the detector integration time. Any thermal source of light filtered with a narrow bandpass filter will fit this model, but so would a standard laboratory laser, which although relatively 'pure', shows random phase variations on timescales of less than a microsecond. These random variations in the wave amplitude cause the interference effects between different sources to be 'washed out' because they are uncorrelated between sources.

This can be shown by revisiting the expression for the intensity of the combination of two beams in Equation (1.7), but considering the case where Ψ_1 and Ψ_2 can have amplitudes and phases that vary randomly with time. In this case of the light pattern at the detector surface will be given by

$$i(x) = \left\langle |\Psi_1(x)|^2 \right\rangle + \left\langle |\Psi_2(x)|^2 \right\rangle + 2\left\langle \mathrm{Re}\left[\Psi_1(x)\Psi_2^*(x) \right] \right\rangle, \qquad (1.23)$$

where the angle brackets denote averaging over the exposure time of the detector. Writing

$$\Psi_1 = \Psi_a e^{-2\pi i \nu \tau_a(x)} \qquad (1.24)$$

and

$$\Psi_2 = \Psi_b e^{-2\pi i \nu \tau_b(x)} \qquad (1.25)$$

where Ψ_a and Ψ_b are the randomly varying complex amplitudes of the incoming waves measured at a fixed point external to the interferometer and $\tau_a(x)$ and $\tau_b(x)$ represent time delays which are fixed during the exposure but can be different for different values of x. This gives an expression for the intensity

$$i(x) = \left\langle |\Psi_a|^2 \right\rangle + \left\langle |\Psi_b|^2 \right\rangle + 2\mathrm{Re}\left[\left\langle \Psi_a \Psi_b^* \right\rangle e^{2\pi i \nu [\tau_a(x) - \tau_b(x)]} \right]. \qquad (1.26)$$

The first two terms are independent of x and so represent uniform illuminations corresponding to the intensity contributions from beams 1 and 2. They consist of time averages of positive quantities, and so are always finite and positive. The third term is a cross-term representing the effects of interference between the beams.

In the case where each beam arises from different stars, Ψ_a and Ψ_b will be uncorrelated in both amplitude and phase, since the variations are due to the spontaneous emission events in atoms in different stars. As a result $\Psi_a \Psi_b^*$ will have a phase which varies randomly during the exposure and so this term will average to zero. The intensity pattern will therefore simply be the sum of the intensity patterns that would be observed if each star were present individually and will contain no cross-term between stars. This can be explained in terms of constructive and destructive interference between the star beams occurring randomly and with equal probability during the exposure time: the average of this will be as if no interference occurred at all.

If instead the two beams come from the same star, then $\Psi_a = \Psi_b$ and so $\Psi_a \Psi_b^* = |\Psi_a|^2$, which is a randomly varying but always positive quantity. This does not average to zero, and gives

$$i(x) = 2\left\langle |\Psi_a|^2 \right\rangle \left[1 + \mathrm{Re}\left[e^{2\pi i \nu [\tau_a(x) - \tau_b(x)]} \right] \right], \qquad (1.27)$$

which recovers the same form for the interference pattern as given in Equation (1.16) by defining $F_0 = \left\langle |\Psi_a|^2 \right\rangle$.

The combined fringe pattern

The previous subsection showed that the fringe intensity pattern seen when observing two close stars will be given by the sum of the fringe patterns from

the two stars individually. In Section 1.4.2 it was shown that the fringe pattern
of the second star will have a phase shift of $-2\pi u\Delta\theta$ where u is the scaled and
projected baseline defined in Equation (1.22), and so the intensity pattern on
the detector will be given by

$$i(x) = F_a\left(1 + \mathrm{Re}\left\{e^{2\pi isx}\right\}\right) + F_b\left(1 + \mathrm{Re}\left\{e^{2\pi i(\Delta\theta u + sx)}\right\}\right) \qquad (1.28)$$

$$= F_a + F_b + \mathrm{Re}\left\{\left(F_a + F_b e^{2\pi i\Delta\theta u}\right)e^{2\pi isx}\right\} \qquad (1.29)$$

where F_a and F_b are the respective fluxes of the two stars. A factor of
2 in the definition of $i(x)$ has been dropped between Equation (1.16) and
Equation (1.28) in order to simplify the mathematics; in any case, the absolute intensity of the fringe pattern is less important than the relative intensities
of the various components of the pattern.

Equation (1.29) shows that the intensity pattern consists of a position-
independent 'DC' term (in analogy to the distinction between DC and AC
in electrical systems) $F_a + F_b$ and a position-dependent sinusoidal term
at frequency s whose amplitude and phase depend on a complex factor
$\left(F_a + F_b e^{2\pi i\Delta\theta u}\right)$. The DC term therefore depends only on the fluxes of the two
stars, while the sinusoidal term depends both on the fluxes of the two stars and,
importantly, on their angular separation $\Delta\theta$.

This dependence on separation can be illustrated by considering the case
where the fluxes of the two stars are identical, i.e. $F_a = F_b$. If the angular
separation of the stars is small so that $\Delta\theta u \ll 1$ then the amplitude of the
sinusoidal fringe will be equal to the DC level, and strong fringes will be seen.
If instead the stars are separated by an angular separation such that $\Delta\theta = 1/(2u)$
then the fringe patterns corresponding to the individual stars will be 180° out
of phase with one another and the amplitude of the summed fringe will be zero.
In other words, the maxima of one fringe pattern will overlap with the minima
of the other and, since the stars are of equal intensity, the fringes will vanish.

1.4.4 Relationships between the fringe pattern and source structure

The preceding analysis of the fringe pattern seen for a pair of stars illustrates a
number of key facts about the interferometer:

- The appearance of the interference pattern is sensitive to the *angular structure* of the object under study: there is an observable difference in the
properties of the fringe patterns seen for a single star of flux $F_a + F_b$ and
a pair of stars with the same total flux, providing the stars are appropriately
spaced.

- The angular scale of the structure to which the experiment is sensitive is of order λ/B radians when the source is overhead.
- Comparing this with Equation (1.1), it can be seen that the angular scales to which an interferometer is sensitive are comparable to the angular resolution of a telescope of diameter B, but B can easily be hundreds of metres, much larger than the size of any existing or planned telescope. For $B = 100\,\text{m}$ and a wavelength $\lambda = 500\,\text{nm}$, λ/B is approximately 1 milliarcsecond.

The last of these points represents the key advantage of interferometers: we can get the angular resolving power of a telescope of size B by using the interference of light from two small collectors separated by a distance B rather than by building a single large collector of size B. This means that B can be extended to sizes larger than the diameter of any feasible single telescope so that previously unattainable angular resolutions can be achieved.

The following section extends the analysis to objects of arbitrary shape, but the above results will be found to be a good guide to the basic features of interferometry.

1.4.5 The fringe pattern from an arbitrary object Vector formulation

The interferometric observation of most interest is one where the object being observed has an arbitrary angular structure. The two-dimensional angular structure of the emission from an object (this can be thought of as what the object 'looks like' from the point of view of a perfect observer) can be characterised by the *object brightness distribution* denoted as $I(\hat{S})$, where $\hat{S}$ is a unit vector representing a particular direction as seen from the position of an observer and where the flux coming from within a small solid angle $d\Omega$ of a direction $\hat{S}$ is given by $I(\hat{S})\,d\Omega$.

The external delay difference for a point source at location $\hat{S}$ given in Equation (1.19) can be written in vector notation as

$$\tau_{\text{ext},12} = \boldsymbol{B}_{12} \cdot \hat{S}/c, \tag{1.30}$$

where $\boldsymbol{B}_{12}$ is the baseline vector, as shown in Figure 1.9. If the delay line is adjusted to give zero OPD for a phase centre in direction $\hat{S}_0$, then

$$\tau_{\text{int},12} = -\boldsymbol{B}_{12} \cdot \hat{S}_0/c, \tag{1.31}$$

and so the net delay for light beams arriving at the beam combiner from a star in direction $\hat{S}$ is given by

$$\tau_{12} = \boldsymbol{B}_{12} \cdot \boldsymbol{\sigma}/c \tag{1.32}$$

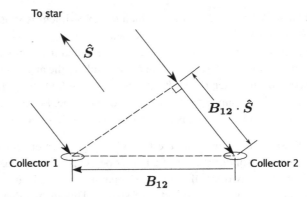

Figure 1.9 Geometry of the paths travelled by light beams from a distant star to two light collectors. The drawing is in the plane containing the vector baseline between the two collectors $\boldsymbol{B}_{12}$ and the unit vector $\hat{\boldsymbol{S}}$ pointing towards the star.

where $\sigma = \hat{\boldsymbol{S}} - \hat{\boldsymbol{S}}_0$ is the offset between the direction to the point source and that to the phase centre. The phase shift of the fringes generated by such a source is therefore given by

$$\phi_{12} = 2\pi \boldsymbol{u} \cdot \sigma \tag{1.33}$$

where $\boldsymbol{u} = \boldsymbol{B}_{12}/\lambda$ is the vector baseline in units of the wavelength.

The (u, v) plane

It is conventional to represent the star and baseline vectors in a right-handed coordinate system with the z axis pointing towards the phase centre, the x axis running towards the east and the y axis running towards the north as shown in Figure 1.10. Writing $\boldsymbol{u} = (u, v, w)$ and $\sigma = (l, m, n)$ gives

$$\boldsymbol{u} \cdot \sigma = ul + vm + nw. \tag{1.34}$$

This expression can be simplified for small fields of view such that $l, m \ll 1$. This is almost always the case in optical interferometry, where the field of view is usually a fraction of an arcsecond so $l, m \lesssim 10^{-6}$. Since $\hat{\boldsymbol{S}}$ and $\hat{\boldsymbol{S}}_0$ both lie on the surface of a unit sphere, then

$$n \approx \frac{1}{2}(l^2 + m^2) \ll |\sigma| \tag{1.35}$$

and so

$$\boldsymbol{u} \cdot \sigma \approx ul + vm. \tag{1.36}$$

From here on, the z coordinates of both the baseline and the source position will be dropped, writing $\boldsymbol{u} = (u, v)$ and $\sigma = (l, m)$, so that $\boldsymbol{u}$ and σ represent

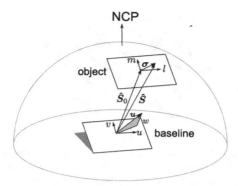

Figure 1.10 Three-dimensional geometry for an interferometric observation, showing the vector baseline $\boldsymbol{u} = \boldsymbol{B}/\lambda$, the phase centre of the image $\hat{S}_0$ and the offset σ. The (u, v, w) coordinate system of the baseline and the (l, m, n) coordinate system for the object are shown (the n coordinate is too small to be visible). The m coordinate points towards the north celestial pole (NCP).

projections on the plane perpendicular to $\hat{S}_0$ of the baseline vector and the sky coordinates respectively. These projected coordinates are said to lie in the '(u, v) plane' (sometimes known as the 'aperture plane' or the 'Fourier plane') and the 'tangent plane', respectively.

The interferometric measurement equation

An object with a brightness distribution $I(\sigma)$ can be represented by a grid of point sources of light spaced by small distances dl and dm with the flux from the point source at position σ being given by $I(\sigma)\,dl\,dm$. As in the case of the pair of sources, each point source gives rise to its own fringe pattern and so the total intensity is the sum over these fringe patterns. In the limit of an infinitely fine grid so that $dl \to 0$ and $dm \to 0$, this sum can be expressed as an integral and the fringe pattern intensity is given by

$$i(x) = \iint_{-\infty}^{\infty} I(\sigma) \left(1 + \mathrm{Re}\left\{e^{-2\pi i \sigma \cdot u} e^{2\pi i s x}\right\}\right) dl\,dm \tag{1.37}$$

$$= 2 \iint_{-\infty}^{\infty} I(\sigma)\,dl\,dm + \mathrm{Re}\left\{e^{2\pi i s x} \iint_{-\infty}^{\infty} I(\sigma) e^{-2\pi i \sigma \cdot u}\,dl\,dm\right\}, \tag{1.38}$$

where the integration limits have been taken to infinity on the assumption that $I(\sigma)$ falls to zero outside some compact region.

The integral can be simplified by defining a complex quantity called the *coherent flux* $F(\boldsymbol{u})$ given by

$$F(\boldsymbol{u}) = \iint_{-\infty}^{\infty} I(\sigma) e^{-2\pi i \sigma \cdot u}\,dl\,dm. \tag{1.39}$$

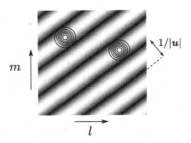

Figure 1.11 Greyscale representation of a two-dimensional sine wave at spatial
frequency u overlaid with a contour map of the brightness distribution of a binary
star. The coherent flux is an integral of the product of the sine wave and the
brightness distribution and so will 'pick' up a binary star with this separation.

Noting that the total flux from the object can be written in terms of a coherent
flux on a baseline of zero length

$$F(0) = \iint_{-\infty}^{\infty} I(\sigma)\, dl\, dm \qquad (1.40)$$

(hence the term *zero-spacing flux*), then the integral can be written

$$i(x) = F(0) + \mathrm{Re}[F(u)e^{2\pi i s x}]. \qquad (1.41)$$

Examination of Equation (1.41) shows that the coherent flux controls three
measurable properties of the fringe pattern: the zero-spacing flux $F(0)$ controls
the average ('DC') level of illumination, the modulus of $F(u)$ determines the
amplitude of the fringe modulation, and the argument of $F(u)$ determines the
phase shift of the fringes with respect to some reference point on the detector.

At the same time, the coherent flux depends on the angular structure of the
object brightness distribution $I(\sigma)$: Equation (1.39) shows that it is an integral across the field of view of the object brightness multiplied by a complex
sinusoid. The real and imaginary parts of the complex sinusoid are a cosine
and sine wave respectively. Each of these oscillates between positive and neg-
ative values over an angular distance on the sky of $1/|u|$ radians as shown in
Figure 1.11, and so the integral will have a large magnitude for objects which
have structure on this angular scale.

Another qualitative example of how the object structure can affect the integral is shown in Figure 1.12. The integral will tend to cancel out for an object
which is smooth on an angular scale of $1/|u|$ while it will tend to be higher for
an object with the same flux which is smaller than this scale. Thus the coherent
flux can be thought of as the flux of the object, filtered to 'pick out' structure
in the object on angular scales of order $\lambda/B_{\text{projected}}$ radians.

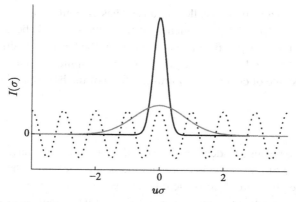

Figure 1.12 Two one-dimensional Gaussian brightness distributions plotted as a function of angular coordinate σ together with a cosine wave at frequency u (shown as a dotted line). The total flux $F(0)$ of both Gaussians is the same but the coherent flux $F(u)$ of the narrower Gaussian at frequency u is greater.

The coherent flux is the key quantity linking the fringe pattern and the object brightness distribution: by measuring the amplitude and phase of the fringes formed between two telescopes, the coherent flux can be determined, and this in turn can be related to the object structure on angular scales, which are inversely proportional to the projected baseline. This relationship, and how it can be used to reconstruct an image of the object, is explored further in Section 2.1.

1.5 Spatial coherence

Interferometry is often explained as being a means to measure the 'coherence' of a light beam. The preceeding analysis has required the use of the concept of coherence only in assuming that the light from two different stars is perfectly incoherent. A brief introduction to the idea of partial coherence is helpful for linking the concepts discussed here to the ideas of coherence, as well as explaining the origin of the term 'coherent flux'.

Given two quasi-monochromatic light beams with complex wave amplitudes Ψ_1 and Ψ_2, the *mutual intensity* of the two beams is defined as

$$M_{12} = \langle \Psi_1 \Psi_2^* \rangle, \tag{1.42}$$

where the angle brackets denote averaging over a time which is long compared to the random fluctuations of the complex wave amplitudes.

The mutual intensity is so called because it has the units of intensity. If $\Psi_1 = \Psi_2$ then $M_{12} = M_{11} = \langle |\Psi_1|^2 \rangle$, which is the intensity of a single beam. The ratio of the mutual intensity to the geometric mean of the beam intensities yields a measure of the correlation of the complex wave amplitudes. This measure is called the 'degree of coherence' (Zernike, 1938) of the beams, and is given by

$$C_{12} = \frac{M_{12}}{\sqrt{M_{11}M_{12}}}. \tag{1.43}$$

When the two beams are identical or perfectly correlated, the magnitude of the mutual degree of coherence is unity and the beams are said to be 'coherent', whereas when they are uncorrelated the degree of coherence is zero and they are said to be 'incoherent'. Intermediate values of the magnitude of the coherence correspond to 'partial coherence'. The degree of coherence and mutual intensity are complex quantities; their phases serve as measures of the mean phase difference between the two beams.

The van Cittert–Zernike theorem (van Cittert, 1934; Zernike, 1938) relates the mutual intensity to the properties of the object emitting the light. In the case of a distant object, the van Cittert–Zernike theorem gives the same result for the mutual intensity of an object with a given angular brightness distribution as is given for the coherent flux of the fringes in Equation (1.39), providing that the light emission from different parts of the object is assumed to be incoherent.

This is not unexpected. The equation for the interference pattern from a pair of quasi-monochromatic beams (Equation (1.26)) can be combined with Equations (1.15) and (1.17) to give

$$i(x) = \langle |\Psi_a|^2 \rangle + \langle |\Psi_b|^2 \rangle + 2\mathrm{Re}\left[\langle \Psi_a \Psi_b^* \rangle e^{-2\pi i s x} \right]. \tag{1.44}$$

The mutual intensity of two beams appears as the complex coefficient of the interference term, i.e. the coherent flux of the fringes. An astronomical interferometer can therefore be considered as a device for measuring the coherence of light beams as a function of their transverse separation across a wavefront, otherwise known as the 'transverse coherence' or 'spatial coherence' of the light.

It should be noted that the terms 'coherent' and 'incoherent' and related terms are used in contexts other than that of the mutual degree of coherence. Coherence is more generally used to indicate phase stability in some signal, for example the 'coherence time' of the seeing is used in Section 3.1 to refer to the time over which the atmospheric perturbations can be considered to be stable, and 'coherent integration' is used in Section 8.6 to indicate integration under conditions where the phase of the fringes can be considered to be stable.

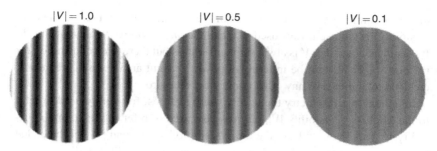

Figure 1.13 Fringe patterns with the same average brightness but different visibilities.

1.6 Nomenclature

At this juncture, a small diversion is necessary in order to establish some nomenclature, as this is somewhat confused for historical reasons. When using the human eye as a detector, the most easily observable property of interference patterns is the *contrast* of the fringes, i.e. the ratio of the intensity modulation of the pattern to the average intensity. The higher this contrast the more easy the fringes can be seen as can be appreciated from Figure 1.13. Michelson (1920) defined the *fringe visibility* as

$$V_{\text{Michelson}} = \frac{i_{\max} - i_{\min}}{i_{\max} + i_{\min}} \tag{1.45}$$

where $i_{\max}$ and $i_{\min}$ are the maximum and minimum intensities in the fringe pattern, respectively. Interestingly, the human psychovisual system (i.e. the combination of the eye and the brain's image-processing systems) is an efficient fringe-sensing tool: under optimum conditions, humans can detect the presence of fringes when $V_{\text{Michelson}} \gtrsim 0.01$.

For a fringe pattern given by Equation (1.41), the Michelson fringe visibility can be written as the modulus of a complex quantity V

$$V_{\text{Michelson}} = |V|, \tag{1.46}$$

where V is a normalised coherent flux

$$V = \frac{F(u)}{F(0)}, \tag{1.47}$$

known as the *complex visibility*. The complex visibility extends the Michelson visibility to include information about the phase of the fringes as well as their contrast.

In radio interferometry the normalisation factor $F(0)$ is often difficult to measure because the high sky background levels mean that determining the

total flux coming from the entire field of view is difficult. As a result, the term
'visibility' is almost always used to mean the *un-normalised* coherent flux F.
It should be noted that V is always dimensionless but there is no convention for
the units of F: it could be in terms of power per unit area per unit frequency,
received total power or any other convenient measure.

In optical interferometry the term visibility is most frequently used to refer
to the Michelson visibility $|V|$, but it is also used to refer to the complex vis-
ibility V. This book will use the term 'visibility' for V, and $|V|$ will be called
the *'fringe contrast'*, the *'visibility modulus'* or the *'visibility amplitude'*.

There is no universally accepted term for F. In this book, the term *coherent
flux* will be used. Other terms commonly used in the literature are *correlated
flux*, *mutual intensity* and *mutual coherence function*. The terms *coherent flux
modulus* or *coherent flux amplitude* will be used for $|F|$.

Both F and V contain essentially the same information, but have different advantages depending on the application. The coherent flux is generally
more useful mathematically as it has the important property of being a linear
function of the object brightness distribution: the coherent flux for an object
composed of two sub-objects is simply the sum of the coherent fluxes for the
sub-objects taken individually.

In contrast(!), the visibility is a non-linear function of the object brightness
distribution, but has the useful property that it allows some constants of pro-
portionality to be discarded. For example, using Equations (1.47), (1.39) and
(1.40) the visibility can be expressed in terms of the properties of the object
under study as

$$V(\boldsymbol{u}) = \iint_{-\infty}^{\infty} I'(\boldsymbol{\sigma}) e^{-2\pi i \boldsymbol{\sigma} \cdot \boldsymbol{u}} \, dl \, dm, \qquad (1.48)$$

where I' is a normalised brightness distribution given by

$$I'(\boldsymbol{\sigma}) = \frac{I(\boldsymbol{\sigma})}{\int_{-\infty}^{\infty} I(\boldsymbol{\sigma}) \, dl \, dm}. \qquad (1.49)$$

Thus the visibility of objects of the same shape but different brightnesses is the
same. The visibility modulus is always unity on baselines short enough that
the object is unresolved.

The coherent flux and visibility can each be used to describe either an
observable property of the fringes or a property of the object. For an ideal
interferometer the two are identical, but for many instrumental reasons the visibility observed in the fringe pattern in a real interferometer may differ from the
visibility that would be computed from Equation (1.49). The visibility in the
latter role will be denoted as the *'object visibility'* and in the former role as the
'fringe visibility'. To distinguish the two mathematically, the fringe visibility

will typically be given subscripts denoting the two telescopes being interfered, e.g. V_{12}, while the object visibility will be denoted as a function of spatial frequency $V(u)$ in analogy to the brightness distribution of the object $I(\sigma)$. The same convention will be applied to the coherent flux.

1.7 Polychromatic interferometry

Previous sections have assumed an interferometer operating at a single wavelength (or more correctly a 'quasi-monochromatic' system, where the range of wavelengths is extremely small). While this ideal can be approached by placing a narrow-band filter in front of the detector, most interferometers observe fringes using a relatively wide spectral bandpass in order to increase the number of photons collected, and this can cause the fringe pattern to deviate from the pattern for a quasi-monochromatic beam.

The fringe intensity pattern at a single frequency v can be written as

$$i(\tau, v) \propto F(0, v) + \mathrm{Re}\left\{F(u, v)e^{2\pi i v \tau}\right\}, \qquad (1.50)$$

where τ is the delay difference between the interfering beams at location x on the detector (typically $\tau \propto x$) and $F(u, v)$ is the coherent flux as a function of the projected and scaled baseline u in a narrow bandpass centred at frequency v (note that for a fixed baseline B, u will be proportional to v). The spectral coherent flux $F(u, v)$ can be understood conceptually as the product of two terms:

$$F(u, v) = V(u, v)F(0, v). \qquad (1.51)$$

The visibility $V(u, v)$ captures the shape of the object at a given frequency, while the zero-spacing flux $F(0, v)$ captures the spectral distribution of the flux from the object, any wavelength-dependent absorption effects of the optics (e. g. narrow-band filters), and the variation of the sensitivity of the detector with wavelength.

To a good approximation, the fringe pattern seen when light containing a range of different wavelengths falls on a detector is the superposition of the fringe patterns seen at each of the constituent wavelengths individually – we can assume that there is no 'cross-interference' between different wavelengths, as the phase difference between light at different wavelengths changes by thousands or millions of radians during the exposure time. The pattern seen in broadband light will be therefore be given by

$$i(x) = \int_0^\infty 2F(0, v) + 2\mathrm{Re}\left\{F(u, v)e^{2\pi i v \tau}\right\} dv. \qquad (1.52)$$

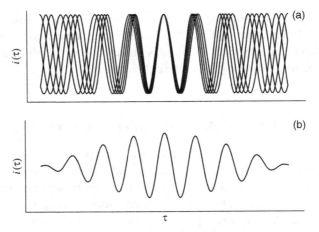

Figure 1.14 Point-source fringe patterns at multiple wavelengths (a) and their sum (b).

The effect of this superposition of fringe patterns can be appreciated by considering a number of simple cases. The first case is a single broadband point source at the phase centre so that the fringe visibility is unity at all wavelengths and baselines. The fringes at different wavelengths will all have a peak at the centre of the detector, but have different frequencies on the detector so that the fringes become more and more out of phase away from the centre. Where the fringe maximum at one wavelength overlaps a fringe minimum at another wavelength, the fringe contrast is diminished, so the fringe contrast tends to fall off away from the centre of the detector as shown in Figure 1.14.

1.7.1 Fourier relationship to the spectrum

A more quantitative insight into the shape of the fringe pattern can be obtained using a Fourier transform. The Fourier transform is discussed in more detail in Section 2.1 and Appendix A; this section can be skipped at first reading by readers less familiar with the Fourier transform.

Equation (1.52) can be rearranged to give

$$i(\tau) = 2 \int_0^\infty F(0, v) \, dv + 2 \int_0^\infty \text{Re} \left\{ F(\boldsymbol{u}, v) e^{2\pi i v \tau} \right\} dv \tag{1.53}$$

$$= \int_{-\infty}^\infty F(0, v) \, dv + \int_{-\infty}^\infty F(\boldsymbol{u}, v) \, e^{2\pi i v \tau} \, dv \tag{1.54}$$

$$= I_0 + I(\tau), \tag{1.55}$$

where $F(u, v)$ has been symmetrised about $v = 0$ such that $F(u, -v) = F^*(u, v)$. The constant I_0 is an offset in intensity corresponding to the flux summed over all wavelengths,

$$I_0 = \int_{-\infty}^{\infty} F(0, v) \, dv, \tag{1.56}$$

and the fringe modulation term $I(\tau)$ is a Fourier transform of the spectral coherent flux,

$$I(\tau) = \mathcal{F} \{F(u, v)\}, \tag{1.57}$$

where it should be noted that the Fourier transform is a one-dimensional transform taken over the frequency coordinate v and not the spatial frequency u.

1.7.2 Coherence length

If the zero-spacing spectrum consists of a flat-topped bandpass with band edges at $v_0 \pm \Delta v$ and the visibility is the same at all wavelengths within this bandpass, then the symmetrised coherent flux spectrum can be represented as a convolution of a 'top-hat' function and a pair of delta functions:

$$F(u, v) \propto \text{rect}(v/\Delta v) * [\delta(v - v_0) + \delta(v + v_0)], \tag{1.58}$$

where the rect function is defined by

$$\text{rect}(t) \equiv \begin{cases} 0 & \text{if } |t| > \frac{1}{2} \\ \frac{1}{2} & \text{if } |t| = \frac{1}{2} \\ 1 & \text{if } |t| < \frac{1}{2}. \end{cases} \tag{1.59}$$

The convolution theorem can then be used to show that the modulation pattern $i(\tau)$ is the product of a sinusoidal fringe pattern with frequency v_0 and a sinc function 'envelope':

$$I(\tau) \propto \cos(2\pi i v_0 \tau) \text{sinc}(\pi \Delta v \tau), \tag{1.60}$$

where the sinc function is defined by

$$\text{sinc}(x) \equiv \frac{\sin(x)}{x}. \tag{1.61}$$

The envelope function goes to zero at

$$\tau = \pm(\Delta v)^{-1}. \tag{1.62}$$

This scaling of the width of the 'fringe packet' as the inverse of the bandwidth is illustrated in Figure 1.15 and is consistent with the expectation that

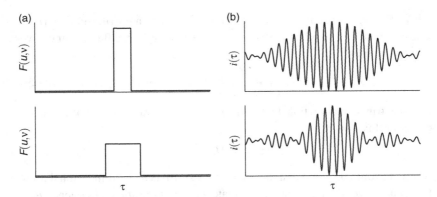

Figure 1.15 Spectral intensity patterns (a) and fringe patterns (b) for light with top-hat spectral bandpasses. The fringe envelope is a narrower, i.e. the coherence length is shorter, for the wider spectral bandpass.

the larger the range of wavenumbers, the more rapidly the fringes at different wavelengths go out of phase with increasing delay τ and so the more rapidly the contrast of the pattern made from the superposition of the fringes drops off.

The number of fringes inside the primary lobe of the fringe envelope is equal to the inverse of the fractional bandwidth, whether expressed in frequency or wavelength:

$$N_{\text{fringe}} = 2 \left| \frac{\nu}{\Delta \nu} \right| \approx 2 \left| \frac{\lambda}{\Delta \lambda} \right|. \tag{1.63}$$

The range of OPD $c\tau$ over which high-contrast fringes can be seen is often called the 'coherence length' of the light (this form of coherence is sometimes called 'longitudinal coherence' to distinuish it from 'lateral coherence', i.e. spatial coherence). The delay lines must equalise the pathlengths in the interferometer to an accuracy which is better than the coherence length in order to see high-contrast fringes.

If fringes are observed through a filter with a fractional bandwidth of 1% (for example a 5-nm bandpass at a wavelength of 500 nm), there will be 200 fringes inside the envelope and at a wavelength of 500 nm, this corresponds to a coherence length of order 0.1 mm. If a bandwidth of order an octave is used, for example the whole of the visible region from 400 nm to 700 nm, then the coherence envelope is of the order of one fringe in size; the position of zero OPD is then uniquely identifiable as the location of the high-contrast 'white-light fringe'.

1.7.3 Bandwidth smearing and field of view

If the object consists of a combination of a point source at the phase centre and one at an angular offset of σ, then the observed fringe system will consist of the superposition of two sets of fringes with envelopes ('fringe packets') which are offset from one another in delay space. If the offset is large enough that the two packets do not overlap, then there will be no position in the fringe pattern where the fringes from both point sources are simultaneously present. If this occurs, then the simple relationship between the fringe visibility and the object visibility will break down – the effects of these errors in the visibility on the reconstructed image are known as 'bandwidth smearing' in radio interferometry.

This will occur when the differential delay between the stars $\sigma \cdot u \lambda / c$ exceeds the size of the fringe envelope as given in Equation (1.62), in other words,

$$\sigma \cdot u \gtrsim \frac{\nu_0}{\Delta \nu}, \tag{1.64}$$

where ν_0 is the centre frequency and $\Delta \nu$ is the bandwidth of the radiation being interfered. The two point sources can then be said to be no longer within the same 'bandwidth-smearing field of view'.

The size of this field of view along a direction parallel to the baseline is given by

$$\Delta \theta_{\text{FOV}} \sim \frac{\nu_0}{\Delta \nu} \frac{1}{|u|} \tag{1.65}$$

and since the angular resolution is given approximately by $|u|^{-1}$ then the field of view can be written as

$$\Delta \theta_{\text{FOV}} \sim \Delta \theta_{\text{res}} \frac{\nu_0}{\Delta \nu}, \tag{1.66}$$

where $\Delta \theta_{\text{res}}$ is the angular size of a resolution element ('resel'). Thus if fringes are observed on multiple baselines using a fractional optical bandwidth of 1% one can in principle make an image up to about 100×100 resels in size before bandwidth-smearing effects become overwhelming.

An alternative approach to making use of a wide spectral bandwidth is to observe fringes in narrow spectral channels but to observe many such channels simultaneously – so-called 'spectro-interferometry'. This approach is more complex but allows the fringe envelope to extend over wider ranges of OPD and therefore larger fields of view. More is said about spectro-interferometry in Section 4.7.4.

1.8 Chromatic dispersion and group delay

In the above analysis, the delays experienced by the light signal have all been assumed to be independent of wavelength. In a real interferometer, the light will pass through a number of optical elements such as vacuum windows and lenses whose refractive index depends on wavelength (they are said to show *chromatic dispersion*). If the differential delay between the light paths travelled by the two interefering beams depends on wavelength then the phase of the fringes will be wavelength-dependent. If the phase shifts by a large amount within the spectral bandpass used to observe the fringes, this could cause the summed fringe across the bandpass to 'wash out'.

The effects of dispersive optical elements within the interferometer can be cancelled by balancing the dispersion in both arms of the interferometer, typically by using identical optical elements in both arms, or by inserting 'compensating plates' of glass. Nevertheless, there will be some residual differential dispersion due to manufacturing tolerances. In addition, in interferometers which use delay lines in air rather than in vacuum, there will be a component of the dispersion which is due to the air in the delay line. This component is time-variable and therefore harder to compensate for, so its effects must be considered.

1.8.1 Atmospheric dispersion

The fact that air in the delay lines causes an unbalanced dispersion can be somewhat counterintuitive, as the aim of the delay lines is to provide a balancing delay to any net external delays, which in a ground-based interferometer consist in part of optical paths in air. Figure 1.16 shows that, providing that the interferometer collecting elements lie in a horizontal plane and assuming a plane-parallel atmosphere, all the optical paths in air external to the interferometer are matched between telescopes. In contrast, the geometric delay due to the phase centre not being directly overhead is a pure vacuum delay.

The beams therefore experience no differential air path due to geometric path effects if the delay line is in vacuum. If the delay line is in air, however, then tens or hundreds of metres of vacuum delay are compensated by a similar amount of air delay, and so the refractive index variations of the air become important and need to be considered. The amount of air path depends on the location of the phase centre, so the effects of dispersion are variable with time.

Light travelling through a distance l in a medium such as the air with refractive index n experiences a delay given by

$$\tau = nl/c. \tag{1.67}$$

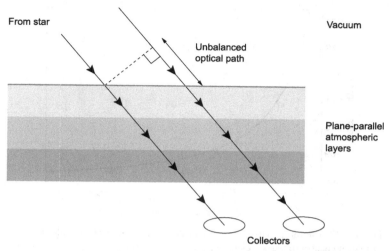

Figure 1.16 Geometry of the external light paths in a plane-parallel atmosphere and a horizontal interferometer, showing that all the optical paths in air are balanced.

If this delay is compensated by an equal path in air in the opposite arm of the interferometer, then the OPD will be given by

$$c\tau_{12} = (n-1)l. \qquad (1.68)$$

Figure 1.17 shows the refractivity $\mu = n - 1$ of dry air as a function of wavelength. It can be seen that μ changes more rapidly with wavelength at blue wavelengths compared with red and near-infrared wavelengths, and this trend continues towards mid-infrared wavelengths.

Water vapour has a refractivity curve which follows the same trend but with a different slope, so the refractivity curve of air depends on its humidity. The refractive index profiles of water vapour and of carbon dioxide have sharp features near their absorption lines at infrared wavelengths, so the detailed shape of the refractivity curve is more complex than is apparent in the figure. This can be important if narrow-band interferometry is being undertaken (Colavita *et al.*, 2004) but is ignored for the rest of this analysis.

The OPD due to differential dispersion can be compensated at a single wavelength by moving the delay line appropriately, but the differential effects with wavelength can still be quite large: the refractivity of air is measured in parts per million, but for a 100-m differential air path, the OPDs experienced at 600 nm and 800 nm will differ by 200 μm, and so the fringe shifts within a bandpass as small as a nanometre will be significant.

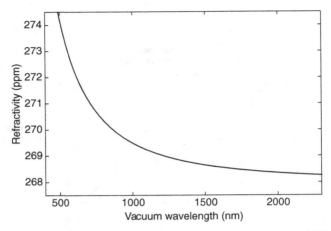

Figure 1.17 The refractivity $\mu = n - 1$ of dry air at 1 atmosphere and 20 °C at visible and near-infrared wavelengths.

This means that if fringes are observed through a filter with a moderate bandpass, they can be washed out by intra-bandpass phase shifts even if the delay lines are moved to cancel the OPD effects of refraction at a single wavelength. It turns out, however, that the fringe contrast will recover if the delay line is offset not by the refractive delay at a given wavelength (the so-called *phase delay*) but by an amount known as the *group delay*.

1.8.2 Group delay

To understand the origin of the group delay, it is helpful to consider a thought experiment in which a symmetrical interferometer consisting of a pair of collectors and a pair of variable delay lines observes a point source directly overhead. If all the optical paths are in vacuum and the delay lines are adjusted so that the net OPD is zero, the phase of the fringes at all wavelengths will be zero. Thus if the fringes are observed using a finite (but small) bandpass the resulting fringe pattern, which is the sum of the fringe patterns at each wavelength within the bandpass, will have both zero phase and high contrast as the fringe maxima will all coincide.

If one delay line is now moved to introduce a (wavelength-independent) delay τ_0 this will cause the fringes to shift in phase. The phase shift as a function of wavelength will be given by

$$\phi(\nu) = 2\pi\tau_0\nu. \tag{1.69}$$

If all the fringe patterns in a finite bandpass are now added together there will be two effects. First, the mean fringe phase will be offset by approximately $2\pi\nu_0\tau_0$, where ν_0 is the central frequency of the bandpass. A second effect is that the fringe contrast will be reduced because of the change of phase within the bandpass – in the worst cases the fringe peak at one wavelength will coincide with the trough at another wavelength within the bandpass.

This can be understood as the effect of the finite-sized fringe envelope due to the optical bandwidth as outlined in Section 1.7: moving the delay line has moved the zero-OPD point and hence the centre of the fringe envelope with respect to the centre of the fringe detector. These two effects, the phase shift of the fringes and contrast reduction, can be seen as the result of the effects of the zero-order and first-order change of phase shift with frequency respectively.

If one delay line is set to introduce a delay of τ_0 and air is then introduced into that delay line then the optical delay introduced by that delay line will be $n(\nu)\tau_0$, where $n(\nu)$ is the refractive index of air at frequency ν. The remaining (i.e. vacuum) delay line can be moved to introduce a delay of $-n(\nu_0)\tau_0$ so that the fringe phase at frequency ν_0 will be zero. The fringes at frequency ν will have a phase given by

$$\phi(\nu) = 2\pi\tau_0\nu\left[n(\nu) - n(\nu_0)\right]. \tag{1.70}$$

The phase variation with ν can be Taylor-expanded about the value at ν_0 to give

$$\phi(\nu_0 + \Delta\nu) \approx 2\pi\tau_0\left[\left.\frac{d(n\nu)}{d\nu}\right|_{\nu_0} - n(\nu_0)\right]\Delta\nu + O(\Delta\nu^2) \tag{1.71}$$

$$= 2\pi\tau_0\nu_0\left.\frac{dn}{d\nu}\right|_{\nu_0}\Delta\nu + O(\Delta\nu^2), \tag{1.72}$$

where $\nu = \nu_0 + \Delta\nu$.

For a sufficiently narrow bandpass the quadratic and higher-order terms in $\Delta\nu$ can be neglected, leaving a linear change of phase with frequency across the bandpass. The linear change of phase with frequency causes a reduction in fringe contrast in exactly the same way as if there had been an OPD error, but the fringe contrast can be restored by adjusting the vacuum delay to introduce a compensating linear phase shift. The required additional delay is given by

$$\tau_{\text{offset}} = \tau_0\nu_0\left.\frac{dn}{d\nu}\right|_{\nu_0}. \tag{1.73}$$

This means that while the compensating vacuum delay needed to make the phase of the fringes zero (known as the *phase delay*) at frequency ν_0 is

$$\tau_{\text{phase}} = \tau_0 n(\nu_0), \tag{1.74}$$

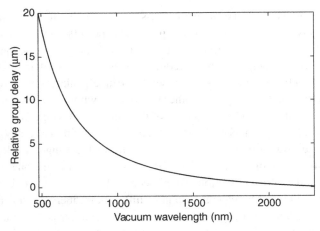

Figure 1.18 The group delay as a function of wavelength for 1 m of dry air at 1 atmosphere and 20 °C. The group delay at 2.5 μm has been subtracted to give a relative group delay.

the compensating vacuum delay for maximum fringe contrast for a bandpass centred at ν_0 (the group delay) is given by

$$\tau_{group} = \tau_{phase} + \tau_{offset} \tag{1.75}$$

$$= \tau_0 \left(n(\nu_0) + \nu_0 \left. \frac{dn}{d\nu} \right|_{\nu_0} \right). \tag{1.76}$$

Figure 1.18 shows the relative group delay introduced per metre of differential air path at different wavelengths. It can be seen that, like the refractive index, the group delay increases sharply towards the blue.

The peaks of fringe packets with different central wavelengths can be separated by significant amounts even for modest amounts of air path difference. For example, with 10 m of differential air path, the fringe envelope centres at wavelengths of 600 nm and 800 nm will be 56 μm apart in OPD. If the width of bandpasses used for forming the fringes at each wavelength are 10% of the respective central wavelength then the fringe envelopes will be 6 μm and 8 μm wide, respectively, and so fringes cannot be seen simultaneously at both wavelengths unless additional differential delay is introduced for each bandpass.

These and other atmospheric dispersion effects can be compensated for either by using narrower bandpasses or by inserting appropriate amounts of glass into the beams in the interferometer so that the dispersion of the glass cancels the dispersion of the air (Tango, 1990; ten Brummelaar, 1995; Lévêque *et al.*, 1996; Davis *et al.*, 1998; Thureau, 2001). A difficulty with the glass compensation approach is that the dispersion of the air is sensitive to temperature and humidity variations, so accurate knowledge of the environmental conditions along the length of the delay line is needed in order to compensate for the dispersion.

2

Basic imaging

2.1 Fourier inversion

A more detailed understanding of the integral defining the coherent flux in Equation (1.39) can be derived by recognising it as a two-dimensional *Fourier transform*, about which there is a well-developed mathematical infrastructure. A brief introduction to the most important properties of the Fourier transform is given in Appendix A.

The Fourier transform can be used to represent any arbitrary (within reason) function f as the sum of a set of sinusoids of different frequencies; for a two-dimensional function $f(x)$ this sum can be written as an integral

$$f(x) = \iint_{-\infty}^{\infty} g(s)e^{2\pi i s \cdot x} \, ds_x \, ds_y, \tag{2.1}$$

where $x = (x, y)$ is a two-dimensional spatial coordinate and $s = (s_x, s_y)$ is a two-dimensional *spatial frequency*. The complex sinusoid $e^{2\pi i s \cdot x}$ is a two-dimensional function whose value is constant along lines perpendicular to the vector s and repeats over a distance $1/|s|$ and so Equation (2.1) is equivalent to saying that a function can be composed by summing a set of sinusoidal 'waves' of different wavelengths and orientations.

Given a function $f(x)$, the coefficients $g(s)$ can be derived using

$$g(s) = \iint_{-\infty}^{\infty} f(x)e^{-2\pi i s \cdot x} \, dx \, dy. \tag{2.2}$$

The process of deriving the coefficient for each spatial frequency is called the Fourier transform and is denoted by the operator $\mathcal{F}$ so that Equation (2.2) is equivalent to writing $g(s) = \mathcal{F}[f(x)]$ and Equation (2.1) is an *inverse Fourier transform* equivalent to writing $f(x) = \mathcal{F}^{-1}[g(s)]$.

With this in mind, Equation (1.39) can be written as

$$F(u) = \mathcal{F}[I(\sigma)], \tag{2.3}$$

36

where u and σ take the places of s and x respectively. Thus if we have measured $F(u)$ we can derive $I(\sigma)$ using

$$I(\sigma) = \mathcal{F}^{-1}\left[F(u)\right]. \tag{2.4}$$

This can straightforwardly be evaluated by computer (and indeed one of the first uses of computers was to evaluate Fourier transforms of data from radio interferometers).

This, then, is the premise of imaging interferometry: the amplitude and phase of the fringe pattern formed by combining the light from two telescopes correspond to the amplitude and phase of a single Fourier component of the angular structure of the object being observed. This component is at a spatial frequency set by the projected baseline u. By changing the vector separation of the telescopes one can sample $F(u)$ at different values of u. With sufficient samples of $F(u)$, an image $I(\sigma)$ of the object under study can be reconstructed using an inverse Fourier transform.

The definition of 'sufficient samples' is the key question, as Equation (2.1) implies that an infinite number of samples is required in theory. This question will be answered in this chapter in three ways. First, an intuitive idea of the kind of information present at different locations in the (u, v) plane will be obtained by looking at the visibility functions of a number of representative objects. Second, strategies for sampling the (u, v) plane will be explored; and finally, an idea of the effects of limited sampling in certain special cases will be obtained, leading to the idea of 'aperture synthesis'.

In all cases, the interferometer will be assumed to be 'phase stable', in other words the measured fringe phase accurately reflects the object coherent phase. This is an adequate model for some radio interferometers, but in Chapter 3 it will be shown that most optical interferometers show strong phase instabilities due to the Earth's atmosphere. Nevertheless, understanding the properties of interferometric imaging under phase-stable conditions can serve as a good basis for understanding the principles of image reconstruction. More details of the processes of image reconstruction in phase-unstable conditions are presented in Chapter 9.

2.2 Visibility functions of simple objects

Before going on to understand in more detail how well images can be reconstructed from interferometric measurements, it is worthwhile examining the characteristics of the visibility for some simple objects so as to get an intuitive 'feel' for the kind of information about an object which is captured by this observable, since the Fourier nature of interferometry is initially quite un-intuitive.

2.2.1 Point source

The simplest object that can be observed is a single point-like source of light, for example a star whose diameter is much smaller than the angular resolution of the interferometer. The properties of a fringe pattern for a point source were derived earlier, but redoing the calculation using the Fourier transform provides an introduction to using the Fourier formalism and interpreting the results.

A point-like object can be represented by a Dirac delta function (delta functions represent impulse-like distributions confined to a single point; their properties are described further in Appendix A):

$$I(\sigma) \propto \delta(\sigma - \sigma_0), \tag{2.5}$$

where σ_0 is the angular coordinate of the source. The Fourier transform of a delta function is a complex exponential, so that

$$F(u) \propto e^{2\pi i u \cdot \sigma_0}, \tag{2.6}$$

and normalising we get

$$V(u) = e^{2\pi i u \cdot \sigma_0}. \tag{2.7}$$

Thus $|V(u, v)| = 1$, in other words the fringe contrast when observing a point source is independent of the length or orientation of the baseline.

2.2.2 Binary star system

The linearity of the Fourier transform means that the coherent flux of the fringe pattern observed for a pair of stars will be the sum of the coherent fluxes for each star observed separately. If both stars are unresolved, and one star with flux F_a is at the phase centre and the second star with flux F_b is at an angular offset of σ_0, then the total coherent flux will be given by

$$F(u) = F_a + F_b e^{2\pi i \sigma_0 \cdot u}. \tag{2.8}$$

The coherent flux can be represented in an Argand diagram as the sum of two vectors as shown in Figure 2.1. The vector of length F_a stays fixed along the real axis while the vector of length F_b rotates in the complex plane as the (u, v) coordinate changes, and so the length of the summed vector will go from $F_a + F_b$ at a maximum to $F_a - F_b$ at a minimum (assuming that $F_a > F_b$).

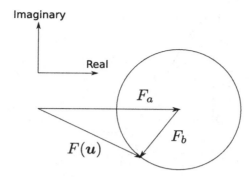

Figure 2.1 Vector sum representation of the coherent flux $F(\boldsymbol{u})$ of a binary system composed of two point sources of flux F_a and F_b.

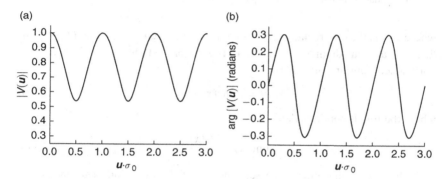

Figure 2.2 Visibility modulus (a) and phase (b) of a binary star with vector separation σ_0 as a function of the normalised projected baseline $\sigma_0 \cdot \boldsymbol{u}$. The flux of the secondary star is 30% of that of the primary star.

The visibility modulus will therefore oscillate between the values of 1 and $(F_a - F_b)/(F_a + F_b)$. The maximum phase excursion will occur when the summed vector is a tangent to the circular locus of the vector of length F_b, giving a maximum phase value of $\arcsin(F_b/F_a)$. The resulting modulus and phase excursions are shown in Figure 2.2.

2.2.3 Uniform disc

A first-order model for the brightness distribution from a nearby star would be a uniform disc a few milliarcseconds across. We can write the brightness distribution for this disc as

$$I(\boldsymbol{\sigma}) \propto \text{rect}(|\boldsymbol{\sigma}|/\theta_d), \tag{2.9}$$

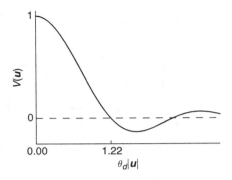

Figure 2.3 Visibility of a point source of a uniform disc such as a star. For a star at the phase centre the visibility is purely real.

where rect is the unit rectangular top-hat function given in Equation (1.59) and θ_d is the angular diameter of the star. The visibility function is given by the normalised Fourier transform

$$V(\boldsymbol{u}) = 2\text{jinc}(\pi\theta_d|\boldsymbol{u}|), \qquad (2.10)$$

where the jinc function is defined as

$$\text{jinc}(x) \equiv \frac{J_1(x)}{x} \qquad (2.11)$$

and where J_1 is the order-1 Bessel function of the first kind. The jinc function, sometimes known as the Besinc function, is like the well-known sinc function (defined in Equation (1.61)), which is the Fourier transform of a one-dimensional top-hat function. The square of the jinc function describes the intensity distribution of the Airy disc pattern, which we met when considering the resolution of a circular telescope.

The visibility function is real everywhere and circularly symmetric in the (u, v) plane and is plotted in Figure 2.3. The visibility modulus generally decreases with increasing projected baseline and goes through a null for $|\boldsymbol{u}| = 1.22/\theta_d$, in other words for a projected baseline length of $1.22\lambda/\theta_d$. Note that the null in visibility does not mean that the measured *intensity* goes to zero: it means that the contrast of the fringes goes to zero so that the fringes disappear and so the fringe pattern becomes a uniformly illuminated field.

Michelson and Pease (1921) used the position of this null in the visibility in their experiment to measure the angular diameter of Betelgeuse. They used a pair of mirrors mounted on a steel beam as 'outriggers' to the Mt Wilson 100-inch (2.5-m) telescope to form the two apertures of the interferometer and observed the fringes seen for different separations of the mirrors.

Noting the mirror separation at which the fringes disappeared gave a measurement of the diameter using $\theta_d = 1.22\lambda/B_{null}$ where B_{null} is the separation of the mirrors when the fringes dissapeared. Betelgeuse has one of the largest apparent diameters of any star at about 40 mas, so that at a visible wavelength of 500 nm the fringes should disappear when the mirror separation is about 3.1 m, approximately the distance observed in the experiment.

A star like the Sun at a distance of 10 parsecs will appear to be about 0.8 milliarcseconds in diameter and the fringes will disappear for baselines of about 700 m when observing at a near-infrared wavelength of 2.2 μm. If the maximum telescope separation available is only $B_{max} = 100$ m then the fringe visibility modulus will always be greater than 99%. We can say that stars of size $\theta_d \ll \lambda/B_{max}$ are 'unresolved'.

2.2.4 Gaussian disc

A uniform disc model is a relatively accurate representation for solar-type stars, but many evolved stars such as Mira variables have extended atmospheres meaning that the edge of the stellar disk is more 'fluffy'. Various limb-darkening models can be used to describe such stars but a simple model of an extreme limb-darkening profile is a two-dimensional Gaussian intensity distribution:

$$I(\sigma) \propto e^{-4\ln 2|\sigma|^2/\theta_d^2}, \tag{2.12}$$

where θ_d is the full width at half maximum (FWHM) of the Gaussian. The Fourier transform of a Gaussian is also a Gaussian; the visibility function is given by

$$V(u) = e^{-4\ln 2|u|^2/\rho_d^2}, \tag{2.13}$$

where $\rho_d = 0.883/\theta_d$ is the FWHM of the visibility curve. The visibility curve is shown in Figure 2.4.

2.2.5 Objects offset from the phase centre

An important set of properties of the Fourier transform, which is useful in deriving visibility functions, are a consequence of the *convolution theorem*. For any two functions $f(x)$ and $g(x)$ with Fourier transforms of and $F(s)$ and $G(s)$, respectively, the convolution theorem states that

$$\mathcal{F}\{fg\} = F * G \tag{2.14}$$

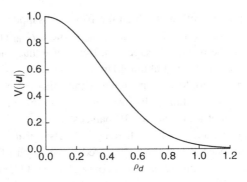

Figure 2.4   Visibility function for a Gaussian disc.

and

$$\mathcal{F}\{f * g\} = FG, \tag{2.15}$$

where $*$ denotes the convolution.

The convolution of two two-dimensional functions $f_1(x)$ and $f_2(x)$ is defined as

$$(f_1 * f_2)(x) \equiv \iint_{\text{All space}} f_1(x')f_2(x - x')\, dA, \tag{2.16}$$

where dA is an element of area. The ideas surrounding convolution are explained in Appendix A, but the simplest useful property of convolution comes from convolution with a Dirac delta function: convolving any function with a delta function offset from the origin by an amount x_0 has the effect of shifting the function by x_0, i.e.,

$$f(x) * \delta(x - x_0) = f(x - x_0). \tag{2.17}$$

This result, together with the convolution theorem, can be used, for example, to compute the visibility function of a star which is offset from the phase centre by some amount σ_0. We write the brightness distribution of the star as

$$I(\sigma) \propto \text{rect}(|\sigma|/\theta_d) * \delta(\sigma - \sigma_0) \tag{2.18}$$

so the convolution theorem means that the visibility function is the product of the visibility functions of the centred disc and the offset delta function:

$$V(u) = \frac{2J_1(\theta_d|u|)}{\theta_d|u|} e^{2\pi i u \cdot \sigma_0}. \tag{2.19}$$

It can be seen from the above expression that the phase of the visibility carries the information about the position of the object with respect to the phase centre, while the modulus carries the information about the size of the object.

2.2.6 Rules of thumb

The above examples display a number of properties typical of visibility functions:

- The visibility at the origin of the (u, v) plane is always unity and the visibility modulus has equal or lower values elsewhere. This is a general property arising from the definition of visibility and the fact that the object brightness distributions are non-negative.
- The smaller the object, the larger the baseline length required to see significant deviations of the visibility from unity: the characteristic baseline length to see these deviations is of the order of λ/θ where θ is a characteristic angular scale for the object. The reciprocal relationship between the baseline length and angular scale arises from the *scaling* property of Fourier transforms (see Appendix A) and is linked to the use of the term 'reciprocal space' to describe the result of a Fourier transform.
- Smoother brightness distributions like the Gaussian have a lower visibility modulus on long baselines than sharper functions like the uniform disc or the binary. This can be thought of as sharp features in the object corresponding to high-frequency structure, and high frequencies are measured on long baselines.
- Brightness distributions such as the uniform disc and the Gaussian, which are symmetric about the phase centre, have visibility functions which are purely real, i. e. the phase is either $0°$ or $180°$ for all $\boldsymbol{u}$. This is a general property of the Fourier transforms of real point-symmetric functions (i. e. functions $f(\sigma)$ whose values are real everywhere and for which $f(\sigma) = f(-\sigma)$ everywhere).
- All the visibility functions have the property of *Hermitian symmetry* in that $V(\boldsymbol{u}) = V^*(-\boldsymbol{u})$, where the star denotes taking the complex conjugate. This is a general property of the Fourier transforms of real functions, but can also be thought of as arising from the fact that the same pair of telescopes p and q can be equally well used to give baselines of $\boldsymbol{B}_{pq}$ or $\boldsymbol{B}_{qp} = -\boldsymbol{B}_{pq}$, the only difference being that, when the roles of the two telescopes are interchanged, the direction of fringe displacement denoted as a positive change of phase needs to be reversed.

As can be seen, these properties of the visibility arise from the properties of the Fourier transform itself. Other important properties of the Fourier transform are summarised in the Appendix A and are mentioned as they arise in the text.

2.3 Sampling the Fourier plane

The coherent flux of the fringes seen in an interferometer consisting of a pair of telescopes provides a sample of the coherent flux at a single location u_{ij} in the (u, v) plane. The number of samples of the coherent flux, in other words the (u, v)-plane coverage, can be increased in a number of ways.

2.3.1 Moving telescopes

The most obvious way to do this is to change the baseline by moving one or both of the telescopes. This was possible to do in the space of a few minutes with the Interféromètre à Z Télescopes (I2T) (Koechlin, 1988) because the telescopes were mounted on rails and had a minimal beam relay system – the beam paths were in open air.

In modern interferometers the beam-relay system is more complex and can take hours to realign. As a result the process of moving a single telescope can take many hours and so baseline reconfiguration is usually done during the day.

2.3.2 Telescope arrays

A faster way to increase the number of samples is to use more than two telescopes and to measure fringes on all the baselines between them. For a set of M telescopes, there are $\frac{1}{2}M(M-1)$ pairs and so in principle this number of (u, v) points can be sampled simultaneously.

Some care has to go into the placement of the telescopes in order to maximise the sampling. Particularly regular spacings of telescopes such as that shown in Figure 2.5(a) have the same baseline repeated a number of times in the array, and so the set of four telescopes allows the sampling of only three distinct baselines instead of six. Such an arrangement is known as a 'redundant' array (lower part of Figure 2.5(b)), while a telescope arrangment such as that shown above in Figure 2.5(b) is called 'non-redundant' and samples six distinct baselines.

Redundancy is less of a problem in two-dimensional array layouts. Figure 2.6 shows one of the layouts of the telescopes in the Magdalena Ridge Observatory interferometer and the corresponding set of baselines. Although this layout has a regular structure and contains repeated baselines, the overall number of distinct baselines is 36, which is 80% of the 45 baselines possible for a completely non-redundant array.

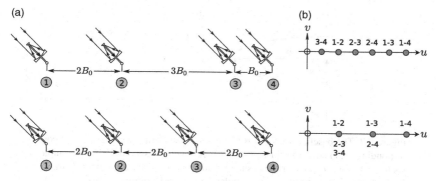

Figure 2.5 Telescopes arranged in different one-dimensional configurations (a) and the corresponding baseline sampling, labeled with the telescope pairs that sample that baseline (b). The lower configuration is termed redundant because at least one baseline is repeated, and as a result fewer distinct (u, v) points are sampled.

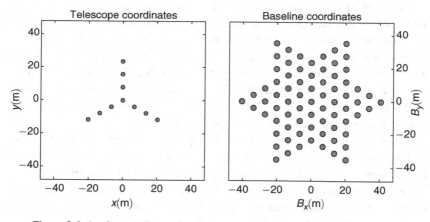

Figure 2.6 A telescope layout for the Magdalena Ridge Observatory interferometer (a) and the resulting baseline coverage (b).

2.3.3 Earth-rotation synthesis

The (u, v) coverage obtainable with a given set of telescopes can be increased by making use of Earth rotation. All ground-based interferometers are sited on a turntable that rotates by $360°$ once per day – the Earth. Since the (u, v) plane is fixed with respect to a coordinate system based on the direction of the object under study, the projection of the baseline between a pair of telescopes which are stationary on the Earth's surface has different projections onto the

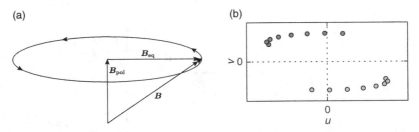

Figure 2.7 The baseline vector B rotating about the Earth's axis (a) and the corresponding (u, v) coverage for an 8-hour track for the observation of an example source (b). The (u, v) coverage is shown with one observation per hour of track. Each observation gives two (u, v) samples due to the Hermitian symmetry of the Fourier transform.

(u, v) plane if observations are made at different times of night as the object rises and sets in the sky.

The effect of Earth rotation is easiest to see for an object at the celestial pole (for example the star Polaris) and an east–west baseline, i. e. one parallel to the equatorial plane of the Earth. In this case the baseline vector describes a circle over the course of 24 hours as seen from the vantage point of the star. Every fringe observation on a given baseline B_{ij} also serves to provide the value of the symmetric baseline B_{ji}, so a set of observations made over a 12-hour period suffice to measure the object visibility function over the complete circumference of a circle in the (u, v) plane.

A more general geometry is shown in Figure 2.7, where the object is not sited at the celestial pole and the baseline vector B has a finite polar component B_{pol} as well as an equatorial component B_{eq}. As the Earth rotates, the polar component remains fixed in space while the equatorial component describes a circle. Seen from the vantage point of a star, the circle is projected on to an ellipse, leading to the baseline describing an ellipse in the (u, v) plane. As shown in Figure 2.7, the centre of the ellipse is offset from the origin of the (u, v) plane by an amount which depends on the projection of B_{pol} onto the plane perpendicular to the line-of-sight to the star.

The combination of using multiple telescopes and Earth-rotation synthesis can lead to good sampling of the (u, v) plane as shown in Figure 2.8.

2.3.4 Wavelength synthesis

If a pair of telescopes is used to make interference fringe observations at a number of different wavelengths, then the (u, v) point sampled at each wavelength will be different as the (u, v) point depends on the ratio of the baseline

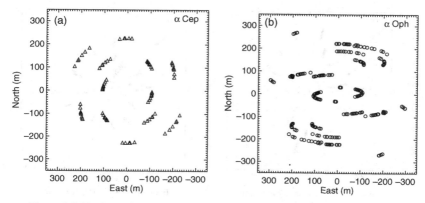

Figure 2.8 Earth-rotation synthesis tracks for the CHARA telescope array observations of two different stars, from Zhao *et al.* (2009). The star α Cephei is relatively close to the celestial pole (declination $\approx 63°$) and so the (u, v) tracks are more circular while α Ophiuchi is nearer to the celestial equator (declination $\approx 13°$) and so the (u, v) tracks are more elongated.

to the wavelength. This can be used to increase the (u, v) sampling of a given set of observations. However, the interpretation of these 'wavelength-synthesis' observations needs to be tempered with caution as they are only easy to interpret if the object shape can be assumed to be the same at all the wavelengths measured. This may be a reasonable constraint if the wavelength range used is small but becomes less valid as the wavelength range increases.

The (u, v) coverage offered by wavelength synthesis consists of a number of scaled copies of the (u, v) coverage available at a single wavelength. Thus the (u, v) coverage will consist of radial streaks as shown in Figure 2.9.

2.4 The image-plane effects of Fourier-plane sampling

2.4.1 The synthesised image

Equation (2.4) shows that it is possible in principle to reconstruct the object brightness distribution exactly if we measure the coherent flux $F(\boldsymbol{u})$ for all values of $\boldsymbol{u}$. In practice we can only measure $F(\boldsymbol{u})$ for at a finite set of discrete values of $\boldsymbol{u}$ rather than completely covering the (u, v) plane.

The effects of this finite sampling can be understood by studying the synthesised image, in other words the image that comes from inverse Fourier

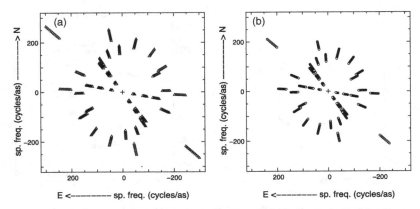

Figure 2.9 Wavelength-synthesis (u, v) coverage of VLTI observations of the star HD87643 in the astronomical H (a) and K (b) bands (from Millour *et al.*, 2009). The radial lines are due to the different spatial frequencies observed on the same baseline at different wavelengths within these bands. Note that the (u, v) coordinates are denoted in units of cycles per arcsecond, a common convention in optical interferometry.

transforming the sampled coherent flux data, and comparing it to the object which produced the data.

For an object brightness distribution given by $I(\sigma)$ and corresponding coherent flux $F(\boldsymbol{u}) = \mathcal{F}[I(\sigma)]$, the sampled coherent flux $\hat{F}$ can be represented by multiplying the true coherent flux by a sampling function consisting of a set of delta functions:

$$\hat{F}(\boldsymbol{u}) = F(\boldsymbol{u}) \sum_k \delta(\boldsymbol{u} - \boldsymbol{u}_k), \tag{2.20}$$

where $\{\boldsymbol{u}_k\}$ is the set of sample locations in the (u, v) plane. A 'synthesised image' (sometimes called the 'dirty image') $\hat{I}(\sigma)$ can be reconstructed by taking the inverse Fourier transform of the sampled data. The convolution theorem can be used to show that

$$\hat{I}(\sigma) = \mathcal{F}^{-1}\left[\hat{F}(\boldsymbol{u})\right]$$
$$= I(\sigma) * b(\sigma), \tag{2.21}$$

where

$$b(\sigma) = \mathcal{F}^{-1}\left[\sum_k \delta(\boldsymbol{u} - \boldsymbol{u}_k)\right]. \tag{2.22}$$

Thus, the effect of making an interferometric image with a given sampling pattern can be described in terms of convolving the object brightness distribution with a 'point-spread function' $b(\sigma)$, which is known in radio astronomy as the 'synthesised beam' or the 'dirty beam'.

The shape of the synthesised beam depends on the details of the (u, v) coverage given by $\{u_k\}$. In general, the angular resolution and field of view of the image will be degraded in comparison with complete coverage of the (u, v) plane, as detailed in the following sections. It is possible to ameliorate some of these degradations by using a process known as 'deconvolution' (discussed in Chapter 9) but the effects of convolution with the synthesised beam can never fully be undone, and so the limitations of the image quality described below give a qualitative idea of the information which is lost as a result of incomplete (u, v) sampling.

2.4.2 Angular resolution

First, we consider an interferometer where the maximum spacing of the telescopes is limited to some value $|B| < B_{max} = \lambda u_{max}$, but the coherent flux has been measured at all spacings less than this. The effect of the incomplete information about $F(u)$ at larger spacings can be modelled by setting the estimated value to zero outside the region where it is known, i. e.

$$\hat{F}(u) = \text{rect}\left(\frac{|u|}{2u_{max}}\right) F(u), \qquad (2.23)$$

where $\hat{F}$ is the estimate of $F(u)$ and the rect function is defined in Equation (1.59). From the convolution theorem (Equation (2.14)) the synthesised image $\hat{I}$ reconstructed by inverse Fourier transforming $\hat{F}$ would then be

$$\hat{I}(\sigma) = I(\sigma) * \text{jinc}(2\pi|\sigma|u_{max}). \qquad (2.24)$$

Thus, the reconstructed image is a convolution of the true image with a jinc function whose angular width (measured from the peak to the first null) is given by

$$\Delta\theta = 1.22/(2u_{max}) = 1.22\lambda/(2B_{max}). \qquad (2.25)$$

The effect of this convolution is to 'blur out' detail in the image on angular scales of order $\Delta\theta$ and hence to degrade the angular resolution. If the object consists of a pair of stars with separation $\Delta\theta$, then the reconstructed image will consist of a pair of jinc functions, where the peak of one overlaps the null of the other. According to the Rayleigh criterion, this pair is the closest pair that can just be resolved, and so the angular resolution of the image is given by $1.22\lambda/(2B_{max})$. Comparing this value with that in Equation (1.1) shows that the angular resolution of an interferometer with maximum baseline B_{max} is the same as that of a single telescope with diameter $2B_{max}$. This gives rise to the term 'aperture synthesis': interferometric measurements can be combined to reconstruct images with angular resolutions comparable to that of a telescope with an aperture of size $2B$, without an aperture of such a size being present.

2.4.3 Field of view

The effect of finite sampling density can be modelled assuming that $F(u)$ is sampled on a regular square grid of points in the (u, v) plane with spacing Δu. The measurements can be represented by

$$\hat{F}(u) = F(u) \sum_{p=-\infty}^{\infty} \sum_{q=-\infty}^{\infty} \delta(u - p\Delta u, v - q\Delta u), \qquad (2.26)$$

with the array of delta functions representing the two-dimensional sampling 'lattice' (note that a lattice is just a two-dimensional version of the Dirac comb function whose Fourier transform is given in Appendix A). The reconstructed image will therefore be a convolution of the true image and the Fourier transform of the lattice, which is a 'reciprocal lattice' with spacing $\Delta \sigma = (\Delta u)^{-1}$, i. e.,

$$\hat{I}(\sigma) = I(\sigma) * \sum_{p=-\infty}^{\infty} \sum_{q=-\infty}^{\infty} \delta(l - p\Delta\sigma, m - q\Delta\sigma). \qquad (2.27)$$

The convolution with each of the delta functions in the reciprocal lattice will produce a shifted image of the original object as shown in Figure 2.10. If the object is small, then it is trivial to remove these 'ghost' images, but if it is greater than $\Delta \sigma$ in size then the ghost images overlap with the true image, and they become difficult or even impossible to disentangle from one another. Thus, good interferometric images can only be made of objects less than this size: the 'interferometric field of view' is therefore set by the density of sampling in the (u, v) plane. Other factors such as bandwidth smearing as discussed in Section 1.7 can also serve to restrict the field of view, but insufficient density of sampling is usually the most important factor that serves to limit the useful field of view.

2.4.4 Information efficiency

The results in the previous two subsections show that if the (u, v) plane is covered by a set of samples at a spacing of Δu and with maximum spacing u_{max}

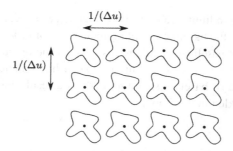

Figure 2.10 The lattice of reconstructed images caused by finite sampling.

then an image with angular resolution of about $1/(2u_{max})$ radians and with a field of view of $1/(\Delta u)$ radians can be faithfully reconstructed. If $u = N \times \Delta u$ then the result could be represented as an image consisting of $2N \times 2N$ independent 'pixels' (these independent resolution elements are more correctly termed resels).

The samples used to make this image consist of an approximately $2N \times 2N$ grid in the (u, v) plane, and each fringe measurement yields two numbers, the real and imaginary parts of the coherent flux, so it might initially seem that twice as many numbers are measured as compared to the independent variables in the image. However, the Hermitian symmetry $F(u) = F^*(-u)$ of the Fourier transform of a real object (see Section 2.2.6 and Appendix A) means that the measurements of coherent flux values in one half of the Fourier plane directly yield values for the opposite half. Hence only half of the number of measurements of the coherent flux need to be made as there are grid points, and so the 'information efficiency' of an appropriately sampled interferometric observation is almost exactly 100%, in the sense that the number of measurements equals the number of unknowns.

2.4.5 Wide-field interferometry

The above developments have assumed that the objects being observed are intrinsically small, specifically that they are smaller than the diffraction limit of a single collector, known as the 'primary beam' in radio interferometry. Radio interferometers can observe fields of view which are larger than the primary beam by using multiple pointings, and this is possible in principle at optical wavelengths. In practice, the possibility of observing such wide fields is restricted by a number of practical issues. The most fundamental of these is the requirement of dense (u, v)-plane sampling. In order to image a field of the size of the diffraction limit of a collector of diameter D, i. e. a field of angular diameter $\theta = \lambda/D$, the baseline sampling required is of order $\Delta B \sim \lambda/\theta = D$; this means that the baseline spacing needs to be of the order of the telescope diameter. No current long-baseline interferometer has sufficient telescopes to approach this kind of baseline sampling and, as a result, wide-field-of-view imaging is not tackled in this book.

3

Atmospheric seeing and its amelioration

The interferometer described in Chapter 1 is an idealised one, where all the optical delays are known and stable. In such an interferometer the phase of the interference fringes measured for a point-like object at a chosen position on the sky (the phase centre) will be zero, and the fringe phase is a good 'observable'. For a number of practical reasons, no existing optical interferometer even approaches this ideal.

First of all, the mechanical tolerances required to know the optical path difference (OPD) internal to the interferometer are formidable. If the position of a single mirror in the optical train is in error by a fraction of a millimetre then the phase of the fringes will be hundreds of radians from the value measured by an ideal interferometer.

While these mechanical errors could in principle be overcome through increasing the precision with which the interferometer is built, there are even more serious phase errors that are introduced external to the interferometer which no amount of increased construction accuracy will overcome. These errors are induced by the passage of starlight through the Earth's atmosphere on its way to the interferometer. The atmosphere introduces phase errors, which are large (many radians) and rapidly varying (on timescales of milliseconds), and so these errors present a fundamental limitation to interferometry from the ground.

The optical effects which cause these rapidly changing phase perturbations are known in astronomy as *seeing*. Seeing can be observed in single telescopes as the 'boiling' of stellar images, which looks similar to the 'shimmering' of images that can be seen on a hot day.

The effects of seeing on interferometers are of a different character to those on a single telescope: whereas on a single telescope the seeing affects the angular resolution which can be obtained, in an interferometer the resolution can be relatively little affected by the presence of seeing but the sensitivity can be dramatically altered. Understanding atmospheric seeing, how it affects interferometers, and how these effects can be ameliorated is the object of this chapter.

3.1 The wavefront perturbation model

The phase delay experienced by a light beam depends on the refractive index of the material it is passing through. The refractive index of air varies with temperature and humidity, so the delay experienced by the light propagating towards the interferometer depends on the detailed temperature and humidity structure of the atmosphere along the two paths.

To first order, the atmosphere is stratified, consisting of a series of horizontal layers. It was shown in Section 1.8 that if the atmosphere's refractive index is uniform within such a set of horizontal layers then the effects of the atmosphere's refractive index along the paths to the collectors of the interferometer cancel one another. However, any horizontal inhomogeneities in refractive index of the atmosphere can introduce disturbances to the OPD and hence to the fringe phase.

The most troublesome refractive index inhomogeneities come from turbulent mixing of atmospheric layers, which can cause random refractive index perturbations, and this will lead to a random perturbation to the delay experienced by the light passing through it. Wind will blow these refractive index perturbations past an observer causing the delay to evolve with time.

The standard model used to describe astronomical seeing therefore has three main ingredients: turbulence, its interaction with the existing gradients of refractive index, and its time evolution. These three ingredients are described to the level required to understand the main impacts of seeing on interferometry; more detailed insight can be gained from numerous papers and books on the subject (see, for example, Fried, 1978; Roddier, 1981; Wheelon, 2001).

3.1.1 Turbulence model

Turbulence in the atmosphere is readily experienced by passengers in an aircraft unlucky enough to fly through it. It can be caused by wind shear at the interface between regions of air moving at different velocities or, near to the ground, by wind striking rocks, trees or buildings. Whatever its cause, the turbulence has a characteristic structure, consisting of random 'eddies' or 'whorls' of moving air, with larger eddies breaking down into smaller and smaller eddies in a fractal type of structure. The statistical properties of this structure under quite general conditions were described by Kolmogorov (1941) and turbulence corresponding to this model is called *Kolmogorov turbulence*.

Turbulence by itself does not cause refractive index variations, as the mechanical energy present in the wind is not large enough to cause significant heating of the air. Instead 'optical turbulence' (that is turbulence which has a significant effect on a propagating beam) occurs only where turbulence and pre-existing gradients of refractive index coincide, for example where turbulence exists at the interface between two layers of air of different temperature.

In such environments, the turbulent motion mixes the air, creating regions of higher and lower refractive index whose structure mimics the structure of the turbulence itself. Figure 3.1 illustrates schematically the effect of these inhomogeneities in refractive index on a beam of starlight. The light arriving at the top of the atmosphere from a star is very close to a plane wave since the star is point-like and very distant. As the light travels through the turbulently mixed air, different parts of the wave are slowed down by different amounts depending on the line of sight through the turbulence, and so the wavefront (which describes a surface across which the light wave has a constant phase at

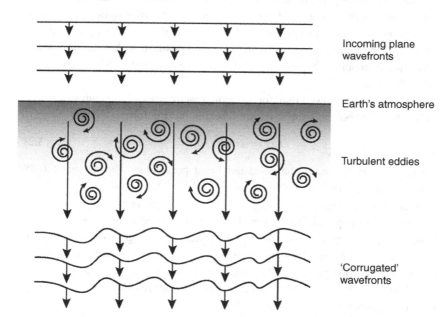

Incoming plane wavefronts

Earth's atmosphere

Turbulent eddies

'Corrugated' wavefronts

Figure 3.1 Schematic diagram of an initially plane wavefront propagating through the atmosphere.

any given instant) changes from being planar to being randomly 'corrugated' by the time the light arrives at a point of observation near the ground.

3.1.2 The spatial structure of the wavefront

The wavefront corrugations are random, and so the spatial structure of the corrugations can only be described statistically. There are a number of complementary ways of describing this structure. One description uses a quantity called the *structure function D_ϕ* of the phase perturbations, which measures how the mean-squared difference in the phase perturbations at two locations on the wavefront varies as a function of their separation. It is defined as

$$D_\phi(r, r') \equiv \left\langle \left| \phi(r' + r) - \phi(r') \right|^2 \right\rangle, \qquad (3.1)$$

where $\phi(r)$ is the value of the perturbation to the phase at a location denoted by the vector r, which is in a plane perpendicular the average direction of propagation of the light beam under consideration, and $\langle \rangle$ denotes taking the average value of a quantity.

If the process that is being described is *homgeneous*, in other words if its statistical properties do not depend on absolute position, only relative position, then it is possible to write $D_\phi(r, r')$ as $D_\phi(r)$. The assumption of statistical homogeneity is a reasonable one for atmospheric turbulence, since in the free atmosphere there is typically no privileged position within the turbulence where the atmosphere can be expected to behave differently than elsewhere in the turbulence.

Tatarski (1961) showed that, for electromagnetic waves propagating through Kolmogorov turbulence, the structure function has the form

$$D_\phi(r) = 6.88(r/r_0)^{5/3}, \qquad (3.2)$$

where $r = |r|$ and r_0 is the so-called Fried parameter. The factor of 6.88 arises from the original definition of r_0 (Fried, 1966), which derives from considerations of the effect of the perturbations on optical imaging systems.

An alternative spatial description of the wavefront perturbations is to decompose the perturbations into sinusoidal ripples or corrugations on different scales. This can be done by taking a two-dimensional Fourier transform of the wavefront corrugations and considering how the amplitude and phase of each Fourier component varies as a function of the wavelength (or equivalently spatial frequency s) of the component.

For any type of homogenous wavefront perturbations, the phases of all the Fourier components will be randomly distributed between 0 and 2π, but the corrugations at different frequencies may have systematically different

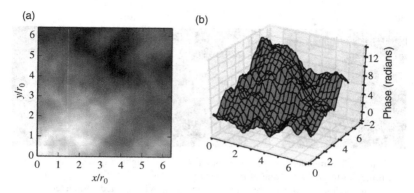

Figure 3.2 Atmospheric wavefront perturbations generated using a numerical simulation. Image (a) is a greyscale image of the surface plotted in (b).

amplitudes from one another. This can be expressed in terms of the *power spectrum* of the ripples, which expresses the mean-squared modulus of a Fourier component as a function of its spatial frequency s. Kolmogorov–Tatarski wavefront perturbations have a power spectrum of the form

$$\Phi(s) = 0.0229 r_0^{-5/3} |s|^{-11/3} . \tag{3.3}$$

It can be seen from this that the low-spatial-frequency (i. e. large-spatial-scale) corrugations have significantly higher amplitudes than their high-frequency counterparts.

The structure function and power spectrum are representations of the same information about the structure of the random wavefront, and one representation can be converted into the other by way of a Fourier transform. For Kolmogorov–Tatarski wavefronts, both representations are isotropic in that they depend only on the magnitude and not on the direction of r or s because it is assumed that the turbulence is itself isotropic.

Both representations show a power-law dependence, which means that the wavefront has a fractal structure, in other words the corrugations on larger scales look like scaled-up versions of the corrugations seen on smaller scales. Figure 3.2 shows a plot of corrugations with the Kolmogorov–Tatarski power spectrum: the similarity in appearance to clouds is not a coincidence as clouds are also caused by turbulence and thus have a similar fractal structure.

3.1.3 Seeing measures

The Fried parameter r_0 is a measure of the 'strength' of the seeing. It denotes a spatial scale across which the phase differences are large enough to have

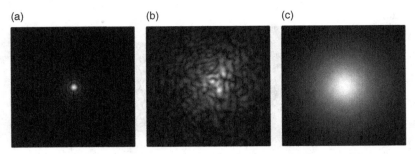

Figure 3.3 Simulated images of a point source of light: (a) seen through a diffraction-limited telescope of diameter d; (b) seen in a short exposure through the same telescope affected by atmospheric seeing with a Fried parameter given by $r_0 = 0.1d$; and (c) seen in a long exposure through the same telescope and seeing.

a significant optical effect. At visible wavelengths and at a good astronomical site r_0 may be about 10 cm in size, so telescopes any larger than this in size will be significantly affected by atmospheric seeing. A point source of light seen through a telescope much larger than r_0 in diameter will appear as a rapidly changing cloud of random 'speckles' similar to that shown in Figure 3.3.

In a long-exposure image, the speckles seen at different instants average out into a broader blur and what is seen is an approximately Gaussian-shaped peak with an angular full width at half maximum (FWHM) of about λ/r_0 – this can be compared with the diffraction-limited spot size of λ/d where d is the diameter of the telescope. The FWHM angular size of a long-exposure image is often used as a measure of atmospheric seeing, and so conditions under which $r_0 = 10$ cm at a wavelength of 500 nm can be described as '1 arcsecond seeing'.

The refractive index of the air is sufficiently constant with wavelength that the perturbations introduced to the optical delay are approximately independent of the wavelength of the radiation passing through it. Since r_0 characterises the spatial scale on which a given magnitude of phase perturbations is seen at a given wavelength, the value of r_0 scales with wavelength as $\lambda^{6/5}$. Thus, for example, the value of r_0 at an infrared wavelength of 2.2 μm will be approximately six times greater than it is at a visible wavelength of 500 nm. The FWHM of a seeing-limited image scales as $\lambda/r_0(\lambda) \propto \lambda^{-1/5}$ and so is a much less strong function of wavelength: 1 arcsecond seeing at 500 nm wavelength corresponds to 0.74 arcseconds seeing at 2.2 μm wavelength.

Seeing is like any other 'weather' condition in that the strength of the seeing can vary from place to place and as a function of time. At a good astronomical site such as Mauna Kea in Hawaii or Paranal in Chile, the median seeing can perhaps be as good as 0.5 arcseconds, corresponding to an r_0 value of 20 cm at a wavelength of 500 nm. Sites with median night-time seeing worse than about 2 arcseconds, corresponding to 5 cm r_0 values, are typically considered unsuitable for siting interferometers.

3.1.4 Temporal seeing

The description of the wavefront perturbations has so far concentrated on the variation of the optical wavefront as a function of lateral position at a single instant in time. To complete the description requires a model of how the wavefront varies as a function of time.

The *frozen turbulence* hypothesis, often called the Taylor hypothesis, assumes that the majority of the temporal variation comes from the bulk motion of the turbulent fluctuations past the point of observation due to the wind. In this scenario, the turbulence is blown past the observer before the structure of the turbulence has time to evolve and, as a result, the temporal evolution of the perturbations to the phase at a given point on the wavefront can be derived straightforwardly from the spatial structure of the perturbations.

The temporal structure function is given by

$$D_\phi(t) \equiv \left\langle \left| \phi(r, t' + t) - \phi(r, t') \right|^2 \right\rangle = (t/t_0)^{5/3}, \tag{3.4}$$

where t_0 is the *coherence time* of the seeing, corresponding to the timescale over which the phase perturbation typically changes by a radian. Note that there are multiple definitions for the coherence time of the seeing (Buscher, 1994; Kellerer and Tokovinin, 2007), but the above definition is perhaps the most common.

If there is only a single layer of turbulence moving at a constant speed v, then the coherence time will be related to the Fried parameter via

$$t_0 = 0.314 r_0 / v. \tag{3.5}$$

For the more realistic case of multiple layers of optical turbulence moving with different speeds and directions, v needs to be replaced by an effective windspeed, which is an appropriately weighted average of the speeds of the different layers. A typical windspeed for an atmospheric turbulence layer might be about 10 m s^{-1}, so that if $r_0 = 10$ cm then $t_0 \approx 3$ ms.

As with the spatial structure, the temporal structure of the perturbations can be represented as a power spectrum, which is given by

$$\Phi(f) = 0.00560 t_0^{-5/3} f^{-8/3}, \tag{3.6}$$

where f is the frequency of the perturbation, meaning that low-frequency oscillations in the phase have much greater amplitudes than those at higher frequencies.

3.1.5 Variation with zenith distance

If a telescope is looking at a star which is not directly overhead, the light received will have travelled through more turbulence between hitting the Earth's atmosphere and reaching the telescope. Thus, the seeing effects will get worse the further the object is from the zenith. Simple geometric arguments (Tatarski, 1961) lead to a prediction for the variation of r_0 with zenith distance z as

$$r_0(z) = r_0(z = 0) \cos(z)^{3/5}. \tag{3.7}$$

The variation of t_0 with zenith distance is more complex, as t_0 depends on the relative orientation of the wind direction and the line of sight in each of the turbulent layers encountered. An experimentally measured dependence (Buscher, 1994) is

$$t_0(z) \propto \cos(z)^{1/5}, \tag{3.8}$$

which is a significantly less severe variation with zenith distance than that for r_0.

3.1.6 Scintillation

If the turbulence is sufficiently far from the point of observation, then diffraction effects can lead to the perturbed optical wavefront having variations in amplitude as well as phase. This gives rise to the phenomenon of stars twinkling when seen with the naked eye. These 'scintillation' effects are usually ignored when modelling interferometers, because under good astronomical seeing conditions the perturbations to the phase of the optical wavefront have a much larger effect on most interferometric measurements than the intensity fluctuations. Neglecting the effects of scintillation is conventionally called the *near-field* approximation (Roddier, 1981).

3.1.7 The outer scale of turbulence

Experimental measurements have confirmed that the Kolmogorov–Tatarski model and the Taylor hypothesis provide good models for the wavefront perturbations seen under astronomical observing conditions on spatial scales from centimetres to metres (Breckinridge, 1976) and on timescales of milliseconds to seconds (Nightingale and Buscher, 1991).

However, the model predicts infinite amounts of wavefront perturbation if extrapolated to infinite scales. This is not surprising, as the Kolmogorov turbulence model only applies for scales that are smaller than the processes which are driving the turbulence.

There is limited evidence as to what this 'outer scale' is in the case of the turbulence causing seeing. The measurements that have been made show that the outer scale is of order 10–100 m in size (Buscher *et al.*, 1995; Davis *et al.*, 1995; Martin *et al.*, 1998; Dali Ali *et al.*, 2010). This means that the Kolmogorov–Tatarski model can be applied on the scale of individual telescopes, but that on the scale of the distances between telescopes in long-baseline interferometers the structure function and power spectrum need to be modified to take account of the effects of the outer scale.

The exact form of modification is unclear, but the evidence is that the structure function 'flattens out' at larger scales so that the rate of increase of phase-difference variance with separation falls below the $r^{5/3}$ power law present on scales less than the outer scale.

3.2 First-order effects on interferometers

The most important effects of the turbulent wavefront perturbations on interferometers can be understood by first considering an interferometer where the individual light collectors are much smaller than r_0 in diameter but the separation of the collectors is much greater than r_0, and where the exposure time used to measure the fringes is much less than t_0. In this case the wavefronts can be considered to have a constant phase across each aperture but the phase will be different between apertures.

The phase of the interference fringes seen in such an interferometer will be offset from the fringes that would be seen in the absence of the atmosphere, by an amount given by $\epsilon_1 - \epsilon_2$ where ϵ_1 and ϵ_2 are the phase perturbations due to the atmosphere over telescopes 1 and 2, respectively. An idea of the typical size of this phase offset can be derived from the root-mean-square (RMS) value. For sources which are nearly overhead this will be given by

$$\sigma_\epsilon = \sqrt{\langle (\epsilon_1 - \epsilon_2)^2 \rangle} = \sqrt{D_\phi(B)}, \qquad (3.9)$$

where B is the length of the baseline. For $B = 10\,\text{m}$ and $r_0 = 10\,\text{cm}$ then, assuming that the outer scale is much larger than 10 m, Equation (3.2) can be

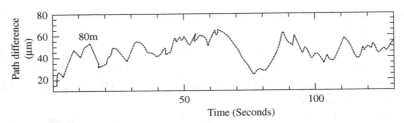

Figure 3.4 The motion of fringes seen on an 80-m baseline on the SUSI interferometer. From Davis *et al.* (1995).

used to derive a value for σ_ϵ of approximately 120 radians, in other words nearly 20 wavelengths or 10 μm of OPD at a wavelength of 0.5 μm. Figure 3.4 shows the measured phase of the fringes seen on the SUSI interferometer showing that OPD fluctuations of many microns are not untypical.

3.2.1 Visibility modulus

The size of the random phase offset introduced by atmospheric perturbations is so large as to render the phase of the fringes meaningless as a measure of the object visibility phase. This is because, even after averaging many measurements, the error is likely to be much greater than a radian. However, the fringe visibility modulus remains unperturbed compared to the no-atmosphere case, providing that the interferometer apertures are much smaller than r_0 and the effective exposure time of the fringe measurement is much less t_0.

Much can be deduced about an object from the visibility modulus alone, and for many years this was the primary observable in optical interferometry. For example, it is possible to measure the diameters and elipticities of single stars, and to measure the separation and brightness differences of binary stars, purely from the changes in visibility modulus as a function of baseline.

Nevertheless, the lack of information related to the phase of the object visibility represents a serious loss of information about the object. For example, any object brightness distribution cannot be distinguished from the same distribution rotated by 180° using visibility-modulus information alone. This is because for any real function $f(x)$, the Fourier modulus is insensitive to reflections about the origin, i. e.

$$|\mathcal{F}\{f(x)\}| = |\mathcal{F}\{f(-x)\}|. \tag{3.10}$$

This clearly can cause problems in the astrophysical interpretation of a measurement, for example in determining the direction of evolution of a binary orbit from a sequence of snapshots of the star locations.

3.2.2 Closure phase

It will be shown in Section 9.6.2 that phase information becomes even more important in complex scenes and is critical to making true images. It turns out that there exists an observable which does capture some object visibility phase information in the presence of large atmospheric disturbances to the phase: this observable is called the closure phase, and was first used to cope with phase instabilities in radio interferometry (Jennison, 1958).

The existence of such an observable can be postulated from considering the amount of phase information measured by an interferometer consisting of N telescopes. If all possible pairwise combinations of the telescopes are made, then there will be $\frac{1}{2}N(N-1)$ independent fringe phase measurements $\{\Phi_{pq}\}$, which are related to the object visibility phases $\{\phi_{pq}\}$ via a set of linear equations of the form

$$\Phi_{pq} = \phi_{pq} + \epsilon_p - \epsilon_q, \tag{3.11}$$

where ϵ_p is the atmospheric perturbation associated with telescope p. There are $N-1$ unknown atmospheric phase perturbations (since it is only the *differences* between the perturbations that matter, one of them can be arbitrarily set to zero) and so if the $\frac{1}{2}N(N-1)$ measurement equations represented by Equation (3.11) are linearly independent (and they are) and $N \geq 3$, then it would seem possible that the equations can be solved for $\frac{1}{2}N(N-1) - (N-1) = \frac{1}{2}(N-1)(N-2)$ linearly independent quantities which are dependent only on the object phases and not on the atmospheric phases.

The simplest such quantity is the sum of the set of measured phases around the closing triangle of baselines formed from any three telescopes in the array. If the three telescopes are labeled p, q and r, it is defined as

$$\Phi_{pqr} = \Phi_{pq} + \Phi_{qr} + \Phi_{rp} \tag{3.12}$$

and is called the *closure phase*.

Substituting Equation (3.11) into Equation (3.12) gives

$$\Phi_{pqr} = \phi_{pq} + \epsilon_p - \epsilon_q + \phi_{qr} + \epsilon_q - \epsilon_r + \phi_{rp} + \epsilon_r - \epsilon_p$$
$$= \phi_{pq} + \phi_{qr} + \phi_{rp}. \tag{3.13}$$

Thus all the atmospheric phase error terms cancel, so we are left with an observable that is dependent only on the object visibility phases. In fact the above analysis can be repeated to show that closure phase is immune to *any* phase error which can be associated with individual collectors (in the nomenclature of radio astronomy, any 'antenna-dependent' phase error), for example any errors in the positions of the mirrors in the beam relay or the delay lines.

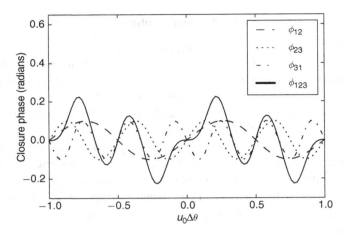

Figure 3.5 The closure phase measured on a binary star system as a function of the separation $\Delta\theta$ of the pair of stars. The flux ratio of the pair is 10:1. In this one-dimensional example, the baselines form a 'linear triangle' with lengths u_0, $2u_0$ and $3u_0$ and the angular separation of the pair of stars is assumed to be parallel to the direction of the baselines. The individual phases ϕ_{12}, ϕ_{23} and ϕ_{31} are shown, together with the closure phase $\phi_{123} = \phi_{12} + \phi_{23} + \phi_{31}$.

As a result using the closure phase eliminates many of the sources of phase error which are otherwise tricky to compensate for.

It is shown in Section 9.6.3 that the closure phase is insensitive to the position of the object with respect to the interferometer phase centre, but it is sensitive to the relative position of features *within* an object. Figure 3.5 shows an example of a binary star measured using a linear 'triangle' of telescopes. An important point illustrated in this example is that the closure phase changes sign if the stars are swapped (i.e. $\Delta\theta$ changes sign). Thus, the closure phase can be used to determine 'which way round' an object is, whereas this would remain ambiguous if only visibility-modulus information were available.

It is shown in Chapter 9 that the closure phase, when combined with the visibility amplitudes, provides sufficient information to robustly reconstruct images in the presence of the atmosphere.

3.2.3 Spectral differential phase

The closure phase is a combination of phases from different baselines, which is independent of the atmospheric phase errors. Another object-dependent but atmospheric-independent phase which can be constructed is called the spectral differential phase (often shortened to 'differential phase'), which uses a combination of phases measured on the same baseline but at different wavelengths.

Consider a two-element interferometer, which measures the fringe phase in N different spectral channels on a single baseline. Instrumental errors will introduce an unknown OPD error of $c\tau_a$ and hence will perturb the fringe phase by $2\pi\tau_a\nu$, where ν is the frequency. As shown in Section 1.8, the atmosphere will perturb the fringe phase by an amount which can be modelled over a limited wavelength range as $\epsilon_0 + 2\pi\tau_b\nu$, where ϵ_0 is a wavelength-independent atmospheric phase error and τ_b is the atmospheric 'group delay'. Thus, the measured fringe phase in spectral channel i will be given by

$$\Phi(\nu_i) = \phi(\nu_i) + \epsilon_0 + \epsilon_1\nu_i, \tag{3.14}$$

where $\phi(\nu_i)$ is the object visibility phase at frequency ν_i, and $\epsilon_1 = 2\pi(\tau_a + \tau_b)$. If the fringes are measured *simultaneously* at three or more wavelengths, then there are more measurements than atmospheric/instrumental unknowns and so an atmosphere-independent measureable can be obtained.

The conceptually simplest such measurement can be derived if fringes are measured at three wavelengths which are equally spaced in frequency such that $\{\nu_i\} = \{\nu_0 - \Delta\nu, \nu_0, \nu_0 + \Delta\nu\}$, where ν_0 is the central frequency and $\Delta\nu$ is the frequency spacing. Subtracting the average fringe phase measured at the outer two wavelengths from the phase measured at the central wavelength yields a quantity in which the atmospheric error terms cancel, leaving the object differential phase:

$$\Phi_{\text{differential}}(\nu_0) \equiv \Phi(\nu_0) - \frac{1}{2}\left[\Phi(\nu_0 - \Delta\nu) + \Phi(\nu_0 + \Delta\nu)\right] \tag{3.15}$$

$$= \phi(\nu_0) - \frac{1}{2}\left[\phi(\nu_0 - \Delta\nu) + \phi(\nu_0 + \Delta\nu)\right] \tag{3.16}$$

$$\equiv \phi_{\text{differential}}(\nu_0). \tag{3.17}$$

This is just one form of the spectral differential phase; with greater numbers of spectral channels, other differential phase measures can be defined. A more sophisticated way of computing a differential phase is to jointly estimate the values of ϵ_0 and ϵ_1 by fitting a phase change which is a linear function of frequency to the data, and attributing any residual higher-order changes of phase with frequency to higher-order variations in the object phase.

The fitting of instrument-introduced spectral phase variations to the measured data can be further extended to include estimation of higher-order terms. For example, atmospheric and instrumental chromatic dispersion (see Section 1.8) will introduce quadratic and higher-order phase variations with frequency. However, if higher-order instrumental phase terms are allowed to be free parameters, then only the object phase variations which are even higher order than the unconstrained instrumental phase terms can be recovered. Thus, only a limited number of instrumental degrees of freedom are fitted to the data.

The interpretation of the object differential phase is simple in the special case where the object is expected to appear as an unresolved point source at the outer wavelengths while the object is expected to appear resolved on the interferometric baselines in question at the central wavelength. This is often because the central wavelength corresponds to a strong spectral emission or absorption line in the object. In this case the differential phase at the central wavelength is simply the object phase at that wavelength, since the object phase at the outer wavelengths is zero.

The differential phase can be used in other situations where there are a priori constraints on the variation of object shape with wavelength. For example, if the object can be assumed to be 'grey', i.e. its shape is independent of wavelength or is a slowly varying function of wavelength, then the differential phase can be used to provide empirical constraints on models of the object.

In many astronomical contexts, however, such assumptions are not appropriate because the variation of the object shape with wavelength cannot be constrained a priori. In such cases the differential phase has seen relatively limited application. This is because in such situations a model for the object shape at every wavelength must be inferred from the data, and the combined model will of necessity have many more parameters than the equivalent single-wavelength model. It is often constrained by many fewer measurements than model parameters, and so the scope for degeneracies, where many widely differing sets of model parameters fit the same data, is greater than in the case of closure phase imaging. Thus, except in a few special cases, imaging is usually obtained using closure phase information; no image reconstruction packages using the differential phase are generally available.

For simplicity, much of the rest of the discussion focusses on the closure phase as the primary source of object phase information for imaging, with the differential phase being reserved for special cases where its unique properties can be best exploited.

3.2.4 The power spectrum and bispectrum

The previous sections have shown that, while the visibility is corrupted by OPD perturbations, the visibility modulus and the closure phase are not (at least for small aperture sizes and for short integration times). These quantities can therefore serve as 'observable' quantities in optical interferometry in place of the complex visibility.

However, these quantities will in general be noisy because of the low light levels involved; therefore, it is advantageous to be able to reduce the noise by taking an average value over multiple independent exposures. Since the exposure time is typically measured in milliseconds, it is possible to take thousands of exposures in a few minutes, and so this averaging (called 'incoherent averaging') can serve to reduce the noise level by more than an order of magnitude. The quantities which are averaged are not the visibility modulus and the closure phase themselves, but rather two quantities from which the modulus and closure phase can be derived. These quantities are the 'power spectrum' and the 'bispectrum', respectively.

The power spectrum is the modulus squared of the fringe coherent flux F_{12}:

$$P_{12} = |F_{12}|^2. \tag{3.18}$$

This can be related to (and in the ideal case is equal to) the power spectrum of the object at spatial frequency u_{12}, where the object power spectrum is given by

$$P(u) = |F(u)|^2. \tag{3.19}$$

The modulus squared captures the magnitude of the coherent flux. It is easier to do analytic computations on the mean of the modulus squared rather than the mean of the modulus: as will be seen in Section 5.5.2, such computations are needed to remove bias terms caused by noise such as photon noise.

The bispectrum is sometimes known as the 'triple product' because it is the product of three coherent fluxes

$$T_{pqr} = F_{pq}F_{qr}F_{rp}, \tag{3.20}$$

where p, q and r denote three different collectors.

The bispectrum of the object can be written

$$T(u_{pq}, u_{qr}, u_{rp}) = F(u_{pq})F(u_{qr})F(u_{rp}), \tag{3.21}$$

and u_{pq} denotes the spatial frequency sampled by the baseline between collectors p and q. Since the relevant baselines form a triangle, then $u_{rp}+u_{pq}+u_{qr} = 0$ and so the bispectrum can be expressed as a function of two spatial frequencies (hence the term 'bispectrum'):

$$T(u_{pq}, u_{qr}) = F(u_{pq})F(u_{qr})F(-u_{pq} - u_{qr}). \tag{3.22}$$

The bispectrum is a complex number whose argument (or phase) is

$$\arg(T_{pqr}) = \phi_{pq} + \phi_{qr} + \phi_{rp}; \tag{3.23}$$

in other words, the closure phase.

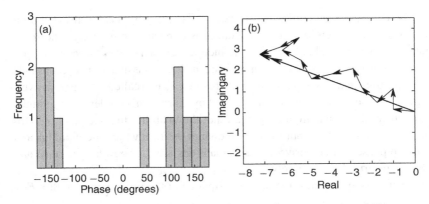

Figure 3.6 Histogram of a sample set of phases (a) and vector averaging of this set of phases (b). The arithmetic average of the phases is $3°$, while the vector average phase is $159°$.

There are a number reasons for averaging the bispectrum rather than the closure phase. The most important of these is that averaging a modulo-360° quantity like the closure phase is fraught with pitfalls, and averaging the bispectrum is a more robust way to obtain an average of such a quantity. This method of averaging is known as a 'vector average' of a complex quantity whose argument is the phase to be averaged.

The advantages of vector averaging can be understood by considering a sample of phase measurements such as those given in Figure 3.6. It can be seen that the samples cluster around a value close to $150°$, but a mean of the sample phase values yields an average phase of about $3°$. This is because around half of the sample phases wrap around from $180°$ to $-180°$, and this biases the average towards zero.

If instead the sample phases are assigned to unit vectors ('phasors') in the complex plane as shown in Figure 3.6(b) then the sum of these phasors is a vector with a phase of $159°$. This vector average is clearly a better central measure of the sample set of phases than the simple mean of the phases.

The argument of the average bispectrum is a form of the vector average. In this case, each phasor has a modulus whose amplitude is the product of the individual fringe amplitudes at the three spatial frequencies. With noisy data, this tends to weight higher-than-average signal-to-noise-ratio data more heavily and so converges faster to the 'true' average phase than a vector average where each phasor has the same length (Woan and Duffett-Smith, 1988).

The modulus of the bispectrum contains information about the magnitude of the object coherent flux at the three component spatial frequencies, but this information is available in a more direct form by measuring the power spectrum at the relevant spatial frequencies, so the bispectrum amplitude information is not as critical to making interferometric images as the bispectrum phase.

3.3 The effects of finite exposure time

If sufficiently small apertures and short exposures are used to make measurements of the fringe pattern, the fringe visibility modulus will be close to the source visibility modulus. If the exposure time is comparable to or larger than t_0 then the fringes will have changed phase during the exposure and so will 'smear out'. As a result, the measured visibility modulus will be lower than the instantaneous fringe visibility modulus by an amount that depends on the fringe motion during the exposure. If the instantaneous fringe intensity pattern has the form

$$I(x,t) = I_0 \left(1 + \mathrm{Re}\left\{ V_0 e^{\mathrm{i}[2\pi sx + \phi_1(t) - \phi_2(t)]} \right\} \right) \qquad (3.24)$$

where V_0 is the object fringe visibility, s is the spatial frequency of the fringes and $\phi_i(t)$ is the fringe phase perturbation over telescope i at time t, then the fringe pattern observed with a finite exposure beginning at time $t = 0$ and ending at time $t = \tau$ is given by

$$I(x,t,\tau) = \int_0^\tau I_0 \left(1 + \mathrm{Re}\left\{ V_0 e^{\mathrm{i}[\phi_1(t) - \phi_2(t)]} e^{2\pi \mathrm{i}sx} \right\} \right) dt \qquad (3.25)$$

$$= \tau I_0 \left(1 + \mathrm{Re}\left\{ V_0 \gamma(\tau) e^{\mathrm{i}[2\pi sx]} \right\} \right). \qquad (3.26)$$

Thus the smeared fringe pattern is a sinusoid with the same fringe frequency s as the unsmeared fringe pattern but a visibility given by $V_0 \gamma(\tau)$, where $\gamma(\tau)$ is a complex *blurring* factor given by

$$\gamma(\tau) = \frac{1}{\tau} \int_0^\tau e^{\mathrm{i}[\Phi_{12}(t)]} \, dt, \qquad (3.27)$$

and where $\Phi_{12}(t) = \phi_1(t) - \phi_2(t)$ is the difference between the phase perturbations at the two telescopes at time t.

3.3.1 Visibility loss for short exposure times

For exposure times such that $\tau \ll t_0$ the phase difference Φ_{12} can be expressed as

$$\Phi_{12}(t) = \Phi_0 + \Delta\Phi(t), \qquad (3.28)$$

where Φ_0 is the average phase for the exposure and $\Delta\Phi(t) \ll 1$ radian. In this case Equation (3.27) can be written

$$\gamma(\tau) = \frac{e^{i\Phi_0}}{\tau} \int_0^\tau e^{i\Delta\Phi(t)} \, dt \qquad (3.29)$$

$$\approx \frac{e^{i\Phi_0}}{\tau} \int_0^\tau (1 + i\Delta\Phi(t) - \frac{1}{2}\Delta\Phi(t)^2) \, dt, \qquad (3.30)$$

where the exponential has been expanded to second order in $\Delta\Phi$. Since Φ_0 is the average phase, then $\int_0^\tau \Delta\Phi(t) \, dt = 0$ and so

$$\gamma(\tau) \approx e^{i\Phi_0} \left(1 - \frac{\sigma(\tau)^2}{2}\right), \qquad (3.31)$$

where $\sigma(\tau)$ is the RMS variation of the fringe phase during an exposure of length τ, in other words

$$\sigma^2(\tau) = \frac{1}{\tau} \int_0^\tau [\Delta\Phi(t)]^2 \, dt. \qquad (3.32)$$

Thus, as might be intuitively expected, the phase of the measured fringe is different from the phase measured in the absence of the atmosphere by an amount given by the average atmospheric phase difference over the exposure. At the same time the modulus of the fringe visibility decreases with increasing variation of the phase during the exposure as $1 - \frac{1}{2}\sigma^2(\tau)$.

It is usually more mathematically convenient to work with the loss in the *squared* visibility modulus, which will be given approximately by

$$|\gamma|^2 \approx 1 - \sigma^2(\tau). \qquad (3.33)$$

This loss will vary from exposure to exposure; the mean loss will be given by

$$\langle |\gamma|^2 \rangle \approx 1 - \langle \sigma^2(\tau) \rangle. \qquad (3.34)$$

An expression of this form appears in a well-known approximation for the intensity of an image formed by an optical system with phase aberrations, known as the Maréchal approximation. The similarity is not accidental, as the calculation to which the Maréchal approximation applies involves integrals similar to that in Equation (3.27).

For longer exposure times where $\langle \sigma^2(\tau) \rangle > 1$, Equation (3.34) is clearly invalid as it predicts a negative value for the squared modulus. A modified version of this expression, which does not have this problem, is

$$\langle |\gamma^2| \rangle \approx e^{-\langle \sigma^2(\tau) \rangle}. \qquad (3.35)$$

This expression agrees with Equation (3.34) in the region of applicability ($\sigma \ll 1$) but gives more realistic answers for the visibility loss for $\sigma > 1$.

The equivalent expression used in aberrated image calculations is known as the 'extended Maréchal approximation' (Mahajan, 1983), and has been found both theoretically (Sandler *et al.*, 1994) and empirically to give an acceptable level of approximation when $\sigma \lesssim 1$.

Assuming a Kolmogorov–Tatarski–Taylor model for the phase perturbations, the mean variance of the phase during an exposure is given by Tango and Twiss (1980):

$$\left\langle \sigma^2(\tau) \right\rangle \approx \left(\frac{\tau}{2.6t_0} \right), \tag{3.36}$$

so the mean-squared visibility loss will vary with exposure time approximately as

$$\left\langle |\gamma(\tau)|^2 \right\rangle \approx e^{-\left(\frac{\tau}{2.6t_0} \right)^{5/3}}. \tag{3.37}$$

3.3.2 The random-walk model for longer exposures

The evolution of the visibility modulus at longer exposure times can instead be derived from a *random-walk* model. In this simplified model, the fringe phase remains roughly constant over some period $\tau_0 \sim t_0$, but changes randomly between successive periods. For an exposure consisting of n such periods, Equation (3.27) simplifies to

$$\gamma(n\tau_0) = \frac{1}{n} \sum_{k=1}^{n} e^{i\Phi_k}, \tag{3.38}$$

where Φ_k is the value of the phase difference $\Phi_{12}(t)$ in the time interval $(k-1)\tau_0$ to $k\tau_0$. Since Φ_k is randomly distributed in the interval 0 to -2π, the summation in Equation 3.38 can be seen as a random walk in the complex plane as shown in Figure 3.7. The standard result for a random walk is that the RMS length of the summed vector is $\sqrt{n}$ times the step size, and therefore $|\gamma|$ decreases as $n^{-1/2}$. As a result the mean-squared visibility loss decreases inversely with the exposure time

$$|\gamma(\tau)|^2 \approx \frac{\tau_0}{\tau}. \tag{3.39}$$

For phase fluctuations given by the Kolmogorov–Tatarski–Taylor model, the mean-squared visibility loss can be evaluated in terms of numerical integrals (Dainty and Greenaway, 1979; Buscher, 1988a) and Figure 3.8 shows the results of these calculations. It can be seen that approximations derived for small and large τ give good predictions for the behaviour in their respective regimes of validity: the extended Maréchal approximation for $\tau \lesssim t_0$ and the

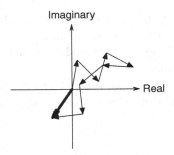

Figure 3.7 A simple random-walk model for the fringe visibility.

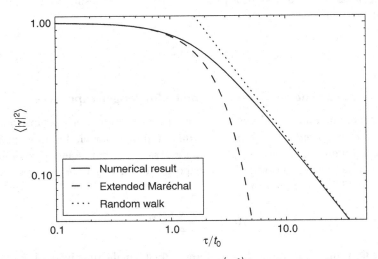

Figure 3.8 The mean-squared visibility loss $\langle |\gamma|^2 \rangle$ as a function of exposure time τ. Also shown are approximate expressions for large and small τ.

random-walk model for $\tau \gg t_0$. In the case of the random-walk model a value of $\tau_0 \approx 1.78 t_0$ gives the best fit.

Visibility calibration

In order to collect enough photons in each exposure, the exposure times used in interferometry are often comparable to or greater than t_0. The loss in visibility due to the finite exposure time can be compensated for by taking many exposures and then normalising the sample RMS visibility using the relevant value given in Figure 3.8. However, there are complications: t_0 is variable and may not be measurable directly; in addition there may be instrumental phase errors which further reduce the visibility. Therefore, the measured visibility is usually calibrated by comparing it with the visibility measured on a nearby reference star with known visibility.

It is desirable that the change in RMS visibility for a given change in t_0 is small so that any changes in t_0 between the observation of the target and observation of the reference star have the least effect on the calibration. This is a function of the gradient of the curve in Figure 3.8 and hence it is desirable to work with short integration times from this point of view. Beyond an exposure time of about $2t_0$, however, a 1% change in t_0 will always give about a 1% change in the mean-squared visibility, as predicted by the random-walk model.

3.4 The effects of finite aperture size

In the previous analysis, the interferometer was modelled as consisting of point-like light collectors. In an interferometer consisting of collectors with sizes comparable with r_0, the effects of the variation in the atmospheric phase perturbations within each collector aperture need to be considered.

For the 'pupil-plane' beam combination scheme used in the example interferometer introduced in Section 1.3, variations in the phase perturbations across the aperture will cause distortions in the fringe pattern. Including the atmospheric phase distortions into Equation (1.16) gives an instantaneous intensity pattern

$$i(x, y) = 2A^2 \left(1 + \text{Re}\left\{V_0 e^{i[2\pi sx + \phi_1(x,y) - \phi_2(x,y)]}\right\}\right) \tag{3.40}$$

$$= 2A^2 \left(1 + \text{Re}\left\{V_0 e^{i(2\pi sx + \Phi_{12}(x,y))}\right\}\right), \tag{3.41}$$

where (x, y) is a two-dimensional coordinate in the illuminated area of the detector, $\phi_i(x, y)$ is the instantaneous phase perturbation due to the atmosphere at coordinate (x, y) within aperture i and Φ denotes a phase difference $\Phi_{12}(x, y) \equiv \phi_1(x, y) - \phi_2(x, y)$. Thus, the positions of the maxima and minima of the fringe pattern shift to reflect the differential phase perturbations at a given location in the aperture as illustrated in Figure 3.9.

These fringe distortions have little effect on the fringe amplitude if the fringe measurement is made locally in small regions in the aperture. However, typically the light level is not high enough to allow accurate measurement of the fringe parameters in small regions, and instead the fringe signal needs to be averaged across the whole aperture. Different deviations to the fringe phase in different parts of the aperture will mean that the averaged fringe will be smeared in a similar way to that experienced in the finite exposure time case.

Figure 3.9 A simulated pupil-plane fringe pattern in the presence of atmospheric phase perturbations.

In Section 8.5 it is shown that there are multiple possible ways to determine the coherent flux from a fringe pattern. One way to do this is to extract the Fourier component in the intensity pattern at the spatial frequency s corresponding to the frequency of the fringes in the unperturbed pattern. This can be achieved by multiplying the intensity by a sinusoid at frequency s and integrating the result over the illuminated region S

$$\hat{F}_{12} = \iint_S e^{-2\pi i s x} i(x, y) \, dx, dy, \tag{3.42}$$

where the circumflex on $\hat{F}_{12}$ denotes an 'estimator' for F_{12}, the fringe coherent flux, and (x, y) is a coordinate in the plane of the detector. Substituting for $i(x, y)$ from Equation (3.40) gives

$$\hat{F}_{12} \approx A^2 V_0 \iint_S e^{i\Phi_{12}(x,y)} \, dx, dy, \tag{3.43}$$

where it has been assumed that $\iint_S e^{2\pi i s x} \, dx, dy \approx 0$ and that $\iint_S e^{i[4\pi s x + \Phi_{12}(x,y)]} \, dx, dy \approx 0$. Both conditions are achieved if the fringe spacing $1/s$ is small compared with the size of the aperture and with r_0.

In the absence of atmospheric perturbations (i.e. when $\Phi_{12}(x, y) = 0$) $\hat{F}_{12} \propto V_0$ and so $\hat{F}_{12}$ provides an estimate of the object coherent flux. Equation (3.43) can be written as

$$\hat{F}_{12} \approx \gamma(D)F_{12}(0),$$ (3.44)

where $F_{12}(0)$ is the coherent flux that would be observed in the absence of the atmosphere, and $\gamma(D)$ is the complex visibility factor introduced by the atmosphere for an interferometer with apertures of diameter D and given by

$$\gamma(D) = \iint_S e^{i\Phi_{12}(x,y)} \, dx, dy.$$ (3.45)

The expression for the visibility factor $\gamma(D)$ due to the atmosphere in Equation (3.45) is of the same form as for the factor $\gamma(\tau)$ due to finite exposure times given in Equation (3.27) but with a two–dimensional integral over the aperture replacing the one–dimensional integral over the integration time. As a result, the techniques developed for calculating the change in visibility due to the temporal atmospheric phase perturbations can be used to derive the change in visibility due to the spatial perturbations.

3.4.1 Visibility loss for small- and large-aperture diameters

In the presence of Kolmogorov–Tatarski turbulent phase fluctuations, the mean phase variance across a single aperture of size D will be given by (Noll, 1976):

$$\langle \sigma^2(D) \rangle = 1.0299 \, (D/r_0)^{5/3} .$$ (3.46)

For two well-separated apertures the wavefront perturbations can be considered to be uncorrelated between apertures and therefore the phase *difference* between the apertures will have a variance twice as large. Using the extended Maréchal approximation given in Equation (3.35) leads to a mean-squared visibility reduction of

$$\langle |\gamma(D)|^2 \rangle \approx e^{-2.06\left(\frac{D}{r_0}\right)^{5/3}}$$ (3.47)

for $D \lesssim r_0$.

In the case of larger apertures, a random-walk model for the phase perturbations from Section 3.3.1 can be used but now assuming that an aperture of diameter $D = nd_0$ can be divided into n^2 subapertures or 'seeing cells' of diameter $d_0 \sim r_0$. Across these cells the phase fluctuations are assumed to be small, but between cells the phases make random jumps of up to 2π radians. For this seeing-cell model, the mean-squared visibility reduction is then given by

$$\langle |\gamma(D)|^2 \rangle \approx \left(\frac{D}{d_0}\right)^{-2} .$$ (3.48)

The mean-squared visibility loss in the general case can be evaluated numerically (Korff, 1973) to give the graph of visibility loss against aperture diameter

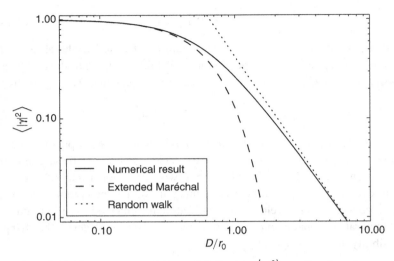

Figure 3.10 The mean-squared fringe visibility loss $\langle|\gamma|^2\rangle$ as a function of aperture diameter D for a pair of well-separated circular apertures in Kolmogorov turbulence. Also shown are approximate forms for $\langle|\gamma|^2\rangle$ when $D \ll r_0$ and $D \gg r_0$.

shown in Figure 3.10, which confirms the applicability of the approximate models for apertures $D \ll r_0$ and $D \gg r_0$. In the latter case a value of $d_0 \approx 0.65r_0$ gives the best fit between the analytical approximation and the numerical integration.

3.5 Adaptive optics

Figure 3.10 shows that the loss in visibility for apertures of only moderate size is appreciable: the RMS visibility decreases by half when the aperture diameter is about r_0, and r_0 is typically a few tens of centimetres while modern astronomical telescopes are typically many metres in size. Thus, the size of apertures which can be used and hence the faintness of the sources which can be observed is limited by the atmosphere and not the technology for building large-aperture telescopes. This limitation can be overcome if the spatial wavefront perturbations across the apertures can be reduced to an acceptable level, effectively increasing the size of r_0. This can be done using *adaptive optics* (AO) systems.

AO systems have become commonplace on large, single telescopes as a way to reduce the effects of atmospheric seeing. Figure 3.11 shows a schematic

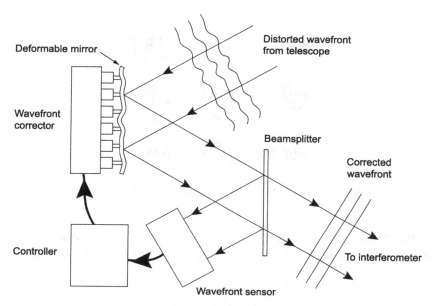

Figure 3.11 Schematic diagram of an AO system on a telescope.

layout for a typical AO system. It consists of a system for measuring the distortions to the incoming wavefronts, called a *wavefront sensor*, a *wavefront corrector*, which is typically a deformable mirror whose shape can be controlled electronically, and a *controller*, which takes the information from the wavefront sensor and attempts to cancel the errors in the wavefront by sending the appropriate signals to the wavefront corrector, causing it to adjust its shape to be the opposite to the atmospheric perturbations. The whole system runs in real time, at update intervals of order t_0 or less, in order to compensate for the distortions before they have time to change.

3.5.1 Wavefront modes

In understanding the degree of correction possible with AO it is helpful to consider the atmospheric wavefront perturbations as being made out of a set of *spatial modes*, that is to say a set of distortion shapes that can be combined to form any possible wavefront perturbation across an aperture. A commonly used set of modes are the Zernike polynomials, which have a number of useful properties. One of these is that many of the lowest-order polynomials correspond closely to well-known aberrations of an optical system, such as coma, spherical aberration and so on.

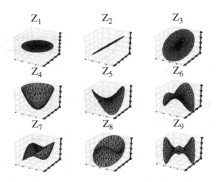

Figure 3.12 Surface plots of the lowest-order Zernike polynomials.

The Zernike polynomials Z_n^m are defined on a unit circle representing the aperture of a telescope and can be separated into radial and azimuthal components

$$Z_n^m(\boldsymbol{r}) = R_n^m(r)f_m(\theta),\qquad\qquad(3.49)$$

where $\boldsymbol{r}$ is a vector inside the circle, (r, θ) are the corresponding polar coordinates, and n and m are integers with $n \in \{0, 1, 2, \ldots\}$ and $m \in \{-n, -n + 2, -n + 4, \ldots n\}$, known as the radial and azimuthal order of the Zernike polynomial, respectively. $R_n^m(r)$ is a polynomial of order n and $f_m(\theta)$ is either $\cos(m\theta)$ or $\sin(|m|\theta)$, depending on the sign of m.

A conventional mapping of the two indices n and m to a single index j has been introduced by Noll (1976). This mapping of groups polynomials with the same radial order n into consecutive values of j such that higher values of j generally correspond to increasing numbers of oscillations within the unit circle.

Figure 3.12 shows a few of the lowest-order Zernike polynomials. The first mode $Z_1 = Z_0^0$ is called the 'piston' mode and corresponds to a uniform phase offset over the telescope aperture. This mode is 'invisible' to a single telescope but piston-mode *differences* between pairs of telescopes are observable by an interferometer as a shift in the phase of the respective fringe pattern. The next two polynomials Z_2 and Z_3 correspond to linear phase ramps across the aperture in orthogonal directions and are conventionally called 'tip' and 'tilt' (which is tip and which is tilt is not well-defined). Changes in these modes correspond to changes in direction of the beam. Higher-order modes correspond to defocus (Z_4), astigmatism (Z_5 and Z_6) and so on.

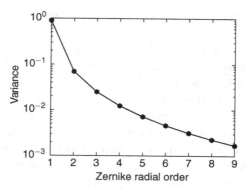

Figure 3.13 The variance of the coefficients of the Zernike modes in the wavefront perturbations produced by Kolmogorov turbulence. The variances for all modes with the same radial order are the same (Noll, 1976); what is plotted is the total wavefront variance contributed by all $n + 1$ modes corresponding to a given radial order n. The variance is plotted for an aperture diameter of $D/r_0 = 1$ and scales as $(D/r_0)^{5/3}$.

Any phase perturbation $\phi(r)$ across a circular aperture can be expressed as a sum of Zernike polynomials with an appropriate set of weights $\{a_k\}$ such that

$$\phi(r) = \sum_{k=1}^{\infty} a_k Z_k(r), \tag{3.50}$$

where $Z_k(r)$ is the Zernike polynomial with index k, which is a function of the two-dimensional position r in the aperture.

The statistical properties of the wavefront perturbations across an aperture can be analysed in terms of the mean-squared values $\{\langle a_k^2 \rangle\}$ of the Zernike coefficients, and for Kolmogorov turbulence this 'spectrum' is heavily weighted towards the lowest-order modes, as shown in Figure 3.13. This reflects the fact that the spatial power spectrum of the Kolmogorov–Tatarski wavefront perturbations (see Equation (3.3)) has more power in the low-frequency corrugations.

In fact, for Kolmogorov turbulence with an infinite outer scale, the theoretical RMS amplitude of the piston mode is infinite. While an infinite outer scale does not exist in practice, infinite-outer-scale models are often used because they are mathematically convenient and represent a worst-case scenario. The infinite amplitude of the piston mode does not present a problem as only the difference of the piston modes between two apertures has any impact on an interferometer, and this difference is finite even if the outer scale is infinite.

Only the non-piston spatial modes are sensed by conventional AO systems – correcting the piston mode requires a fringe tracker, and this is discussed further in Section 3.8. The practical implication of the rapidly falling amplitudes of the Zernike modes with mode order is that, for an AO system with a limited number of degrees of freedom, these degrees of freedom are best used in removing the lowest-order modes of the wavefront, leaving the higher-order modes uncorrected. All modern optical interferometers incorporate at least tip–tilt AO correction at each aperture, because removing just the tip and tilt modes reduces the variance of the wavefront by nearly 90% (Noll, 1976).

3.5.2 Interferometry with AO

Adding an AO system to each of the collecting elements of an interferometer will reduce the level of phase difference variation between the interfering wavefronts and so the integral in Equation (3.43) will be larger than the value obtained in the absence of AO. Thus, AO serves to increase the visibility of the fringes, by an amount which depends on the original level of wavefront distortion and on the level of correction provided by the AO.

The visibility loss due to the residual wavefront aberrations in an interferometer with an AO system can be calculated semi-analytically in some cases (Wang and Markey, 1978; Wilson and Jenkins, 1996) but a more flexible method to calculate these and other quantities that are affected by atmospheric wavefront perturbations is to use numerical simulations. This involves generating random wavefront perturbations across telescope apertures whose statistical properties are the same as those given by Kolmogorov–Tatarski theory, simulating the operation of the AO system in partially correcting these aberrations and simulating the combination of beams from two or more telescopes to form fringes. The simulation is run using a large number of different random realisations of the phase perturbations and the average values of quantities, such as the squared visibility, are computed. Typically a few thousand realisations are averaged to achieve results with accuracies of order a few percent.

The simulated perturbations are usually generated on a square grid representing a sampled wavefront; the sampling is usually at sub-r_0 intervals. One way to generate these *phase screens* is by filtering numerically generated 'white noise' so as to yield phase perturbations with approximately the correct power spectrum (McGlamery, 1976) but there are a host of other techniques which can be deployed. A wavefront perturbation simulation is included in the online material for this book (see Appendix B).

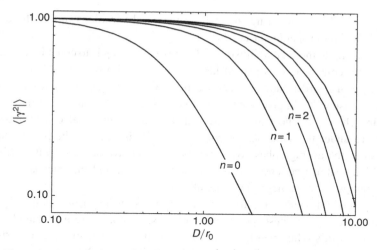

Figure 3.14 The mean-squared visibility loss $\langle|\gamma|^2\rangle$ as a function of aperture diameter D for an interferometer with an AO system at each telescope. AO systems which remove increasing numbers of low-order wavefront modes are shown and are labelled by the maximum Zernike radial order n corrected. An interferometer with no correction ($n = 0$) is also shown for comparison. The graphs for $n = 3 - 5$ are not labelled but follow the sequence of the lower-order graphs.

Simulation of the AO system can be quite complex depending on the level of realism required. For simplicity, it is often assumed that the AO acts as a perfect 'Zernike filter', removing all Zernike polynomials below some given order in a wavefront, and leaving as residual perturbations all the constituent higher-order polynomials.

Using these simulations, the mean-squared visibility loss for an aperture of diameter D, $\langle|\gamma|^2(D)\rangle$, can be calculated for different levels of AO correction. The results from such a simulation are shown in Figure 3.14, where it can be seen that, as expected, AO systems decrease the visibility loss for a given size of telescope, and that the larger the number of modes corrected the larger the telescope can be before the mean-squared visibility loss factor falls below a given value, for example 50%.

3.6 Spatio-temporal effects

The previous sections have examined the effects of finite exposure time on infinitesimal apertures or the effect of finite aperture size with infinitesimal exposure time. In reality, both the exposure time and the aperture size will be finite, and to derive the visibility loss in this case is complex because the frozen-turbulence model means that the temporal variations in the phase are coupled with the spatial variations.

One extreme model that illustrates this cross-coupling is to consider an interferometer consisting of a pair of large telescopes with perfect AO correction of all Zernike orders above piston. In this case the temporal variation of the piston phase will still give rise to fringe smearing over a finite integration time, but the change of the fringe phase will be slower than that of a point-like aperture under the same seeing conditions. This is because the piston phase measures the average phase over the aperture; the averaging operation filters out high-spatial-frequency ripples in the wavefront, and this means that the high-temporal-frequency ripples are also removed.

In practice, however, the fringe smearing for short exposures is not significantly affected by telescope diameter (Tubbs, 2005). There are a number of reasons for this. The first of these is that a significant contributor to the fringe motion is low-frequency but large-amplitude ripples in the phase, which are much less affected by 'aperture filtering'. A second reason is that most AO systems used in interferometers are not perfect correctors and as a result the corrected wavefront contains significant residual high-order modes. Although these modes have zero average phase, they can affect the fringe phase because of the non-linearity of the dependence of the fringe phase on the aperture phase (Buscher *et al.*, 2008). Since they can evolve more rapidly than the piston phase they can contribute significantly to the short-timescale evolution of the fringe phase.

As a rule, therefore, treating the effects of spatial and temporal wavefront fluctuations on fringe visibility separately and multiplying the visibility loss factors for the two gives a reasonable approximation to the visibility loss experienced in practice. To achieve higher-accuracy calculation of the loss requires running a realistic simulation of the atmospheric temporal and spatial variations together with a realistic simulation of the spatio-temporal performance of the AO system.

3.7 Spatial filtering

An alternative method to remove the spatial atmospheric perturbations is to use *spatial filtering*, which in contrast to AO is a passive method that does not require any moving parts. As its name suggests, spatial filtering selectively removes the high-frequency 'noise' in a wavefront caused by the atmosphere, leaving the unaberrated component of the beam behind.

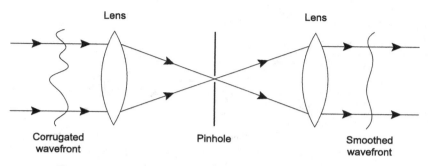

Figure 3.15 Layout of a spatial filtering arrangement using a pinhole.

The simplest form of spatial filter consists of a lens and pinhole arrangement as shown in Figure 3.15: this arrangement is commonly used for removing aberrations from laser beams. The diameter of the pinhole is chosen to be approximately the same as the first lobe of the Airy pattern from an unaberrated beam, i.e. approximately $2.44 f \lambda / D$, where f is the focal length of the lens, λ is the wavelength of the radiation and D is the beam diameter. Aberrations to the beam cause some of the light to be scattered into a halo of speckles around the main lobe, but this light does not pass through the pinhole. At the output of the lens on the other side of the pinhole is therefore a filtered version of the input beam with only the low-frequency components of the input beam retained – the high-frequency components, including the wavefront corrugations, have been removed.

An even more effective form of spatial filtering can be obtained by replacing the pinhole with a length of single-mode optical fibre as shown in Figure 3.16. Optical fibres typically consist of a glass core surrounded by a lower-index glass cladding. Light is confined to the core by total internal reflection and so the fibre acts as an optical waveguide. As with all waveguides, the light propagates as a set of discrete 'guided modes', each of which is characterised by a distinct spatial pattern of the electromagnetic field within the waveguide. Single-mode (or 'monomode') fibres have cores only a few wavelengths across, and can support only one spatial mode (the "fundamental mode") at the wavelength of operation.

Any input beam focussed on the fibre end can be thought of as being the sum of the fundamental mode and higher-order modes, and so a monomode fibre effectively filters out any component of the beam which does not match the fundamental mode; all other modes are either reflected from the fibre entrance or are dissipated in the cladding. As a result, the light coming out of the fibre has a

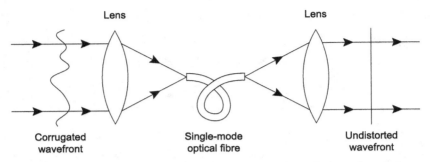

Figure 3.16 Layout of a spatial filtering arrangement using a single-mode optical fibre.

fixed amplitude and phase profile, which are independent of the amplitude and phase profile of the beam that is focussed on the fibre. Thus, a monomode fibre acts as a particularly 'pure' spatial filter, rejecting 100% of the 'bad' aberrated input beam and producing a beam with a fixed profile at the output.

Spatial filtering does not come for free, however, as components of the beam which do not match the filter profile go to waste. In the case of a pinhole filter, any light falling outside the pinhole is lost.

In the case of a monomode fibre, the loss calculation is slightly more complex. If a beam with a transverse amplitude and phase distribution given by the complex field amplitude $\Psi(x, y)$ is focussed by a perfect lens onto a single-mode fibre then the amplitude and phase of the coupled fundamental mode is given by the complex 'modal coefficient',

$$a_0 \propto \iint_S \Psi(x, y)\Psi_0(x, y) \, dx \, dy, \qquad (3.51)$$

where $\Psi_0(x, y)$ is the so-called far-field pattern of the fibre fundamental mode (Wagner and Tomlinson, 1982). This pattern is equivalent to the wavefront which would be observed if a beam in the fundamental mode were to be propagated backwards through the fibre and the lens.

By the Cauchy–Schwarz inequality, the coupling efficiency of the beam into the fibre is therefore a maximum when $\Psi(x, y) \propto \Psi_0(x, y)$ and decreases, the more dissimilar the incoming beam is to the fibre far-field pattern. The far-field pattern typically has a constant phase across the wavefront and so aberrated wavefronts will couple less light into the fibre.

Thus, spatial filtering is best used in combination with adaptive optics, with the AO providing the majority of the wavefront correction and the spatial filter 'cleaning up' the remaining aberrations. Figure 3.17 shows how the coupling efficiency of light from an AO-corrected telescope into a single-mode fibre

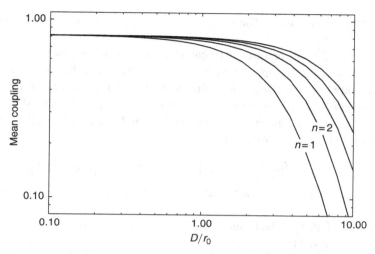

Figure 3.17 Fraction of starlight coupling into a single-mode fibre spatial filter as a function of the size of telescope for different levels of adaptive correction. The maximum Zernike radial order n removed by the adaptive optics system is indicated for each graph. The graphs for $n = 3-5$ are not labelled but follow the sequence of the lower-order graphs. The fibre mode is assumed to have a far-field pattern that is Gaussian in shape and has a $1/e$ radius which is 0.9 times the radius of the beam from the telescope: this choice gives approximately optimal coupling.

varies with aperture size and level of correction. These values were calculated using simulations similar to those described in Section 3.5.2. Even in the absence of atmospheric wavefront perturbations, the maximum coupling efficiency is approximately 80% because the uniformly illuminated beam from a circular telescope aperture does not match the Gaussian fibre far-field pattern (Shaklan and Roddier, 1988).

3.7.1 The effects of fibre coupling fluctuations

If atmospherically corrupted beams from two different telescopes are passed through single-mode fibres and then interfered, the fringe contrast will be high because all the spatial phase distortions in the incoming beams have been filtered out. However, even if the source is a point source (i. e. one with an object visibility modulus of unity), the instantaneous fringe contrast in any single short exposure of the fringes is unlikely to be unity. The reason for this is that the fraction of the incoming light coupling into each fibre is a fluctuating random variable, and so the two beams being interfered will, more likely than not, have different intensities. The mismatch in intensity will cause a reduction in fringe contrast.

This reduction in contrast can be calculated by considering the interference of two beams with complex wave amplitudes $\eta_1 \Psi_0$ and $\eta_2 \Psi_0$ corresponding to the output of two fibres where the inputs both have wave amplitude Ψ_0 (corresponding to a point source at the phase centre) and the intensity coupling efficiencies are given by $|\eta_1|^2$ and $|\eta_2|^2$. From Equation (1.44), the fringe intensity pattern will be given by

$$
\begin{aligned}
i(x) &= |\eta_1|^2 \left\langle |\Psi_0|^2 \right\rangle + |\eta_2|^2 \left\langle |\Psi_0|^2 \right\rangle + 2 \left\langle |\Psi_0|^2 \right\rangle \operatorname{Re} \left[\eta_1 \eta_2^* e^{-2\pi i s x} \right] \\
&= F_0 \left(|\eta_1|^2 + |\eta_2|^2 + 2 \operatorname{Re} \left[\eta_1 \eta_2 e^{-2\pi i s x} \right] \right),
\end{aligned}
\tag{3.52}
$$

where $F_0 = \left\langle |\Psi_0|^2 \right\rangle$ is the incident intensity. The fringe visibility modulus will therefore be given by

$$
|V| = \frac{2|\eta_1||\eta_2|}{|\eta_1|^2 + |\eta_2|^2}.
\tag{3.53}
$$

Writing the ratio of the coupling amplitudes as $f = |\eta_1|/|\eta_2|$ then

$$
|V| = \frac{2}{f + 1/f},
\tag{3.54}
$$

which is unity when $f = 1$ but lower than unity for all other values.

Since the visibility depends only on the coupling ratio $|\eta_1|/|\eta_2|$, then if it is possible to measure this coupling ratio on an exposure-by-exposure basis then the visibility can also be calibrated on an exposure-by-exposure basis. This is known as *photometric calibration* and is discussed further in Section 8.8.

Since the visibility depends on how well matched the instantaneous intensity coupling factors $|\eta_1|^2$ and $|\eta_2|^2$ are, the RMS fringe visibility modulus depends on the level of coupling fluctuations: the larger the fluctuations the lower the RMS visibility will be. However, it is straightforward to show that under relatively general conditions the RMS coherent flux is dependent only on the mean coupling into the fibres and not the exposure-to-exposure fluctuations in the coupling.

Equation (3.52) shows that the coherent flux in a given exposure is given by

$$
F_{12} = 2F_0 \eta_1 \eta_2
\tag{3.55}
$$

and so its mean-squared modulus will be given by

$$
\left\langle |F_{12}|^2 \right\rangle = 4F_0^2 \left\langle |\eta_1||\eta_2|^2 \right\rangle,
\tag{3.56}
$$

where angle brackets are used to denote averaging over multiple exposures. The intensity coupling terms can be written as

$$|\eta_1|^2 = \langle |\eta_1|^2 \rangle + n_1 \qquad (3.57)$$

and

$$|\eta_2|^2 = \langle |\eta_2|^2 \rangle + n_2, \qquad (3.58)$$

where n_1 and n_2 are zero-mean fluctuation terms. In terms of this decomposition, the mean-squared coherent flux modulus will be given by

$$\langle |F_{12}|^2 \rangle = 4F_0^2 \left(\langle |\eta_1|^2 \rangle \langle |\eta_2|^2 \rangle + \langle n_1 \rangle \langle |\eta_2|^2 \rangle + \langle n_2 \rangle \langle |\eta_1|^2 \rangle + \langle n_1 n_2 \rangle \right). \qquad (3.59)$$

Since n_1 and n_2 are by definition zero mean, the second and third terms in the above expression average to zero and, if n_1 and n_2 are uncorrelated, the last term also averages to zero, giving

$$\langle |F_{12}|^2 \rangle = 4F_0^2 \langle |\eta_1|^2 \rangle \langle |\eta_2|^2 \rangle. \qquad (3.60)$$

The assumption of uncorrelated fluctuations is a good approximation for coupling the light from well-separated telescopes into fibres, and under this assumption the mean-squared coherent flux is independent of the coupling fluctuations n_1 and n_2. This means that it may be possible to calibrate the coherent flux by measuring the mean intensity coupling, in an alternative form of photometric calibration. This is explored further in Section 8.8.2.

3.8 Fringe tracking

Conventional adaptive optics works in the domain of individual collectors in an interferometer and cannot measure the piston-mode wavefront perturbations. The form of AO used to correct the piston perturbations in an interferometer is called fringe tracking.

The 'wavefront sensor' in this case is called the 'fringe sensor' as it is some form of beam combiner that uses light from a reference source (which is usually the science target itself) to form fringes between two or more telescopes. Measurements of the fringe patterns are used to derive estimates of the differential piston errors between pairs of telescopes and these estimates are used to adjust internal delays in the interferometer, using, for example, the delay lines, to compensate for the measured errors.

The main beam combiner of the interferometer can be used as the differential piston sensor, but many interferometers use separate beam combiners for making the 'science' measurements of the object visibility and for fringe

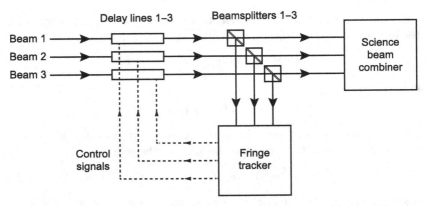

Figure 3.18 Example layout for an interferometer with a separate fringe-tracking beam combiner and science beam combiner.

tracking as shown in Figure 3.18. This is so that the two combiners can be optimised separately: the fringe-tracking combiner is typically optimised for speed and sensitivity at low light levels, while the science combiner can be optimised for fringe measurement precision, baseline coverage, spectral resolution and/or coverage and a host of other factors.

3.8.1 Cophasing and coherencing

Fringe tracking cannot entirely remove the effects of the atmospheric piston, as it is difficult to distinguish phase shifts due to the atmosphere from those due to the intrinsic structure of the reference source used for fringe tracking. Fringe tracking is therefore mainly concerned with reducing the effects of *changes* in fringe phase with time and with wavelength, and in practice the closure phase or spectral differential phase are still used as observables in order to remove 'quasi-static' errors in the visibility phase.

As described in Section 3.3, the change of the differential piston with time causes a visibility loss due to fringe smearing if the fringe exposure time is comparable to or greater than t_0. A fringe tracker can be used to compensate for this change in real time (i.e. on timescales $\ll t_0$) so that the fringes on the science beam combiner are 'frozen'. This freezing of the fringes is called 'cophasing' and means that the detector in the science beam combiner can be read out on timescales much longer than t_0. This allows cameras with long readout times to be used and also allows the effects of readout noise (discussed in Section 5.2.2) to be reduced.

The OPD errors due to the combination of atmospheric and instrumental effects in the interferometer can be of order 100 μm or more. Such a delay causes phase shifts between the fringes at neighbouring wavelengths and, as explained in Section 1.7, phase shifts within a spectral bandpass will cause a reduction in fringe visibility due to the finite size of the fringe envelope.

To take an example, the fringes observed using a bandpass corresponding to the astronomical infrared *J* band with a centre wavelength of 1.25 μm and a fractional bandwidth of 16% will have a fringe envelope which is approximately 8 μm wide. If the OPD error is uncertain at the level of many tens of microns, then most of the time no fringes will be visible. This implies that some form of active method of finding and tracking the coherence envelope is required unless considerably narrower spectral bandpasses are used.

The requirements of such a 'coherencing' system are not as severe as for a cophasing system, since the allowable error in the OPD is of the order of the fringe envelope width rather than being a fraction of a fringe. Since it takes longer for the OPD to change by an envelope width than to change by a fraction of a wavelength, the coherencing system can operate more slowly than a cophasing system. Coherencing methods will be examined in more detail in Section 6.4.

3.9 Wavelength dependence of atmospheric perturbations

In the above analysis it has been implicitly assumed that the OPD changes caused by turbulence are wavelength-independent. In reality, the OPD changes are caused by refractive-index changes in the air and the refractive index of the air changes with wavelength, as discussed in Section 1.8. Thus, atmospheric chromatic dispersion needs to be considered when considering the wavefront perturbations due to seeing.

The OPD changes due to atmospheric seeing are caused by the light traversing regions of differing temperature and humidity. The refractive-index change due to a small temperature offset δT is given approximately by

$$\delta n(\lambda) \approx \mu(\lambda)\frac{\delta T}{T_0}, \tag{3.61}$$

where $\mu = n - 1$ and T_0 is the ambient temperature in kelvins. At optical and near-infrared wavelengths the effects of temperature fluctuations are typically much larger than the effects of humidity fluctuations (Colavita *et al.*, 2004) and so if a delay perturbation of $\tau(\lambda_1)$ is introduced at a wavelength λ_1, then at a wavelength λ_2 the equivalent delay is

$$\tau(\lambda_2) = \frac{\mu(\lambda_2)}{\mu(\lambda_1)}\tau(\lambda_1), \qquad (3.62)$$

and so the difference in delay at the two wavelengths is given by

$$\tau(\lambda_2) - \tau(\lambda_1) = \tau(\lambda_1)\left(\frac{\mu(\lambda_2)}{\mu(\lambda_1)} - 1\right). \qquad (3.63)$$

Figure 1.17 shows that μ changes by about 2% over the visible region of the spectrum, and so this gives rise to a 2% error in the OPD calculated by assuming that the atmosphere is non-dispersive over this wavelength range. This error can be neglected when considering the turbulent fluctuations over an aperture of order r_0 in size, where the phase perturbations at any given wavelength are of order 1 radian, as phase errors of 0.02 radians will typically have a negligible effect on the fringes.

In contrast, the turbulent perturbations to the fringes typically encountered in long-baseline interferometry are of the order of a few tens of microns, and so the error in assuming that the atmosphere is non-dispersive can be as large as a few hundred nanometres of OPD (i. e. several radians of phase) over the same wavelength range. When a fringe tracker is operated at a different wavelength from the science wavelength, it may therefore be necessary to compensate for the difference in phase change seen at one wavelength with that seen at another wavelength using knowledge about the chromatic dispersion properties of the atmosphere.

Problems can still arise because the magnitude of the turbulent water-vapour ('wet') fluctuations can become significant compared to a wavelength even though they are usually less than the temperature-induced 'dry' fluctuations (Colavita et al., 2004). The wavelength dependencies of the wet and dry fluctuations are different, and so assuming that all the perturbations seen by the fringe tracker are 'dry' will lead to an error in the phase correction at the science wavelength.

4

Interferometers in practice

Previous chapters have assumed a rather generalised and abstract interferometer. This chapter looks at the how the functionality of this abstract interferometer is implemented in reality. This exposition will make use of examples from existing interferometers, with the aim of giving an idea of the diversity and ingenuity of the implementations of this functionality.

4.1 Interferometric facilities

The following is a brief summary of the interferometric facilities which were operational at the time of writing or expected to be operational within the next few years. The systems are listed in order of the date (or expected date) of 'first fringes' on each of these interferometers. More information can be found in the online supplementary material (see Appendix B).

Aperture-masking instruments Masking the aperture of a single telescope to convert it into an interferometer was used in the very earliest days of interferometry, and yet it is still a competitive technique for many astronomical measurements (Tuthill, 2012). Because the implementation challenges for aperture masking are in some ways different to those for separated-element interferometry, discussion of the practical features of this technique is deferred until Section 4.10.

SUSI The Sydney University Stellar Interferometer (Davis *et al.*, 1999) is sited near to the radio telescopes of the Australia Array in Narrabri, Australia. It operates at visible wavelengths and has baselines ranging from 5 m to 640 m (currently only baselines up to 80 m have been commissioned).

NPOI The Navy Precision Optical Interferometer (formerly the Navy Prototype Optical Interferometer) (Armstrong *et al.*, 1998) is sited on the

Lowell Observatory Anderson Mesa Station in Arizona, USA. It operates at visible wavelengths and is capable of performing wide-angle astrometric measurements as well as interferometric imaging. It has baselines from 2 m to 437 m (at the time of writing only baselines from 8.8 m to 79 m have been commissioned).

CHARA array The Center for High Angular Resolution Astronomy array (ten Brummelaar *et al.*, 2005) is sited on Mt Wilson, California, USA. It operates at visible and near-infrared wavelengths, and has baselines ranging from 34 to 330 m.

VLTI The Very Large Telescope Interferometer (Schöller, 2007) on Mt Paranal in Chile consists of four 'unit telescopes', which act part-time as independent 'single-dish' telescopes for conventional astronomical observations and four 'auxilliary telescopes', which are used full-time for interferometry. A suite of interferometric instruments allow operation at near-infrared and mid-infrared wavelengths with baselines from 8 m to 200 m. Narrow-angle astrometric observations will be possible with the GRAVITY instrument.

LBT The Large Binocular Telescope is a pair of telescopes on a common mount with a centre-to-centre separation of 14 m, sited on Mt Graham, Arizona, USA. When the beams from the two telescopes are combined, interferometric baselines up to 22 m are available (Angel *et al.*, 1998). The LBT interferometric configuration is in many ways more similar to a single telescope with an unusual aperture than to an interferometer.

MROI The Magdalena Ridge Observatory Interferometer is under construction in New Mexico, USA (Buscher *et al.*, 2013). It will operate at visible and near-infrared wavelengths and will have baselines ranging from 8 m to 340 m.

The discussion will generally follow the progression through the elements of the beam train of a typical optical interferometer. An example beam train, that for the MROI, is shown in Figure 4.1.

4.2 Siting

Any optical telescope needs to be situated on a site with adequate clear night skies. In addition, the performance of an interferometer is strongly dependent on having good atmospheric seeing conditions, as explained in Chapter 3.

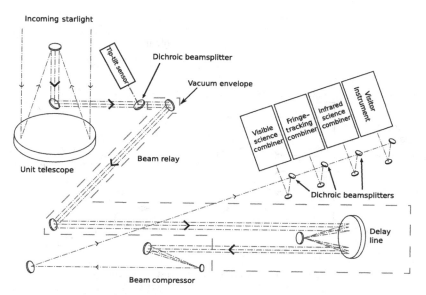

Figure 4.1 The optical beam train for one arm of the MROI.

The only way to completely get rid of the effects of seeing is to put the interferometer above the Earth's atmosphere, i. e. into space. Putting an interferometer into space provides its own formidable challenges, not least of which is cost: the likely cost of any interferometric space mission exceeds the cost of all the telescopes currently on the ground.

The alternative is to place the interferometer on a site with minimal seeing. The sites with the best seeing are typically on the tops of relatively sharply-peaked mountains, as this means that the wind over the site has not previously interacted strongly with the ground and therefore is less turbulent. This conflicts with another desirable characteristic for an interferometer, namely that it should be sited on a large flat area to allow for flexibility in positioning the telescopes. As a result the site chosen for most interferometers is a usually a compromise between these two desiderata.

Two examples of this compromise can be seen in the CHARA array and the NPOI. The CHARA array is situated on Mt Wilson, with excellent seeing, but the available baselines are restricted by the availability of suitable locations in the relatively hilly local topography. The NPOI is situated on the Anderson Mesa, which is quite flat, but although it is at a relatively high altitude the seeing is somewhat worse than at Mt Wilson.

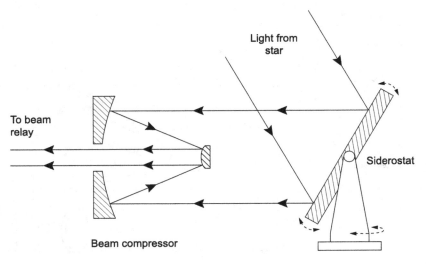

Figure 4.2 Schematic layout of a siderostat light collector.

4.3 Collectors

In radio interferometry, the interferometric system as a whole is often called a 'telescope' with the individual collectors being 'dishes' or 'antennae'. This causes some confusion at optical wavelengths where the collectors are themselves often telescopes, so the individual collectors are sometimes called 'unit telescopes' to distinguish them from the 'synthesis telescope' they form part of. In this book, the collectors will typically be called 'telescopes' and telescope arrays 'interferometers'.

The light-collecting apertures of the array serve to sample the wavefront of light from the target and inject the light into the rest of the system. The simplest collectors are 'siderostats', flat mirrors which can be rotated to reflect light from the selected target in a fixed direction. Siderostats are often followed by 'beam compressors' as shown in Figure 4.2. The beam compressors are fixed telescopes that serve to reduce the diameter of the starlight beam and hence reduce the size of the optics needed in the rest of the system.

An alternative is to use a conventional astronomical telescope to capture the light, and an arrangement of mirrors to send the light out in a fixed direction. Telescopes are preferred over siderostats for larger apertures, because the size of the siderostat mirror needs to be considerably larger than the required collecting area in order to take account of obliquity effects, and the

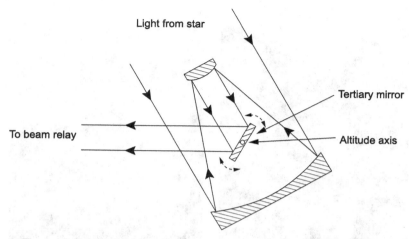

Figure 4.3 Schematic layout of an alt-alt telescope. The tertiary mirror rotates at half the angular rate of the primary-secondary pair in order to keep the output beam fixed.

cost and weight of such mirrors increases rapidly with size. However, using conventional telescopes has its own problems: the moving structure is less compact and so more prone to vibration.

One unusual geometry of telescope that is used in interferometry is the so-called *alt-alt* configuration as shown in Figure 4.3. This arrangement has a rotating tertiary mirror to allow the light to exit the telescope in a fixed direction after only three reflections, the same number of reflections as needed for a siderostat with a beam compressor (indeed, this arrangement can be considered as being equivalent to a beam compressor followed by a siderostat). A more conventional telescope design such as an *alt-az* design can be more compact, but requires seven or more reflections to achieve the same effect, leading to a loss in light and beam quality.

The diameter of the telescope needs to be large in order to collect as much light as possible, consistent with having adequate wavefront quality. Since the main determinant of wavefront quality is atmospheric seeing, the telescopes are typically a few r_0 in size if only tip–tilt correction is applied, but can be many r_0 for telescopes with higher-order AO.

The telescope aperture diameter tends to increase with increasing wavelength since r_0 scales as $\lambda^{6/5}$. This is illustrated by the collectors for SUSI and the VLTI shown in Figure 4.4: SUSI operates at visible wavelengths and has siderostats with an effective diameter of 14 cm, while the VLTI operates in the 1–10 μm range and so has telescopes with a diameter of 1.8 m.

(a)   (b)

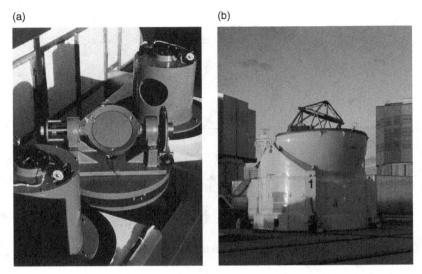

Figure 4.4 Collecting elements: the siderostats on SUSI (a, from Davis *et al.*, 1999) and the alt-az auxilliary telescopes of the VLTI (b, courtesy ESO).

4.3.1 Atmospheric dispersion correction

A number of other subsystems are usually sited at the telescopes. An atmospheric dispersion corrector can be used to make sure that beams at different wavelengths travel in the same direction: it corrects the chromatic refraction effects caused by the 'prism' of atmosphere in the stellar beam which occurs when the target is not directly overhead.

These effects on the tip and tilts of the beams are sometimes called 'angular dispersion' to distinguish them from the 'piston dispersion' or 'linear dispersion' effects caused by there being different amounts of air path between different telescopes and the beam combiner, as discussed in Section 1.8.

Because the atmospheric angular refraction at any given wavelength is the same at all telescopes, it is possible to do without an atmospheric dispersion corrector and still produce high-contrast fringes. The main negative effects of angular dispersion are to do with getting light at all wavelengths through the various apertures in the system. For example, the angular separation between the incoming beams at wavelengths of 500 nm and 800 nm from a target which is 45° from the zenith is about 0.6 arcseconds for a telescope located at an altitude of 2 km. This separation is then multiplied by the angular magnification of the unit telescope: the value of 10 for the VLTI is typical, so the separation of the beams coming out of the telescope is perhaps 6 arcseconds. Over a 200 m path, this corresponds to a relative transverse displacement of the beams at the two wavelengths of about 6 mm, a substantial fraction of the 18-mm pupil size for the VLTI auxilliary telescopes.

Even more severe effects are apparent if spatial filtering is used, as the angular size of the pinholes or fibres used have typical angular sizes of a fraction of an arcsecond when projected on the sky, so if light at one wavelength is centred on the pinhole, light at another wavelength may not pass through at all.

4.3.2 Adaptive optics

All long-baseline interferometers have AO at least at the level of tip–tilt correction and the wavefront-correcting element is typically sited close to the telescope. This is because the largest aberrations must be corrected before they can have a significant effect on the propagation of the light towards the beam combiner. The most common example of this is that tip and tilt errors of many arcseconds (referred to the sky) can arise because of imperfect telescope tracking. If these were corrected at some considerable distance from the telescope then the starlight beams will likely miss any correcting element unless all of the intermediate optics were considerably oversized.

The wavefront-sensing element is often not sited at the telescope: this way aberrations further down the optical train can also be sensed and corrected. However, siting the wavefront sensor at the far end of the optical train from the telescope means that the field of view of the sensor is limited and less light reaches it because of losses in the rest of the beam train.

4.4 Beam relay

4.4.1 Free-space propagation

The signal collected by the unit telescopes must be transferred to a central point for processing and combining with the signal from other telescopes. In all the interferometers described above, this is done by propagating the light as collimated beams, using a set of flat mirrors to direct the beams along an appropriate path. The mirrors are polished to a high accuracy in order to minimise the wavefront distortion induced by the reflections, and have highly reflective coatings in order to minimise the loss of light. These coatings are usually based on metals such as aluminium, silver or gold, and can achieve reflectivities of 98% or more over a wide wavelength range.

4.4.2 Evacuated air paths

The path along which the beam propagates can be hundreds of metres in length, so the effects of air refraction along this path can be significant, causing chromatic dispersion and also seeing. To ameliorate some of the effects of refraction, the air path can be enclosed in ducts, as is done at the VLTI, but more often the beam propagates through evacuated tubes: this latter approach is used at the SUSI, NPOI, CHARA and MROI.

The level of vacuum used in the beam path need not be at the levels needed for cryostats or particle accelerators. These latter scenarios require pressures at the 10^{-8} bar level or lower, while pressures of order a millibar or so are sufficient to ameliorate the worst effects of air refraction in an interferometer. Reducing the air pressure inside the pipe to 1 mbar is equivalent to reducing 100 m of differential path to 10 cm of equivalent air path, and this is sufficient to reduce the differential dispersion effects discussed in Section 1.8 to tolerable levels.

As in normal air, atmospheric seeing can occur in vacuum pipes with residual air, if pockets of hot and cold air occur in the optical beam. We can consider an extreme example of a pipe 400 m long which is filled with random 'blobs' of air heated to about 20 °C above the ambient temperature. This could be due to some source of heat in the pipe, for example a motor operating a delay line.

If each blob is of order 20 cm in diameter and the pressure of the air in the pipe is 1 mbar, then the differential optical path introduced by one such blob of air is about 3.84 nm at optical wavelengths. If the pipe is 50% filled with such random blobs, then a beam of light will intercept about 1000 such blobs. Thus, the root-mean-square (RMS) optical pathlength perturbation will be of order $\sqrt{1000} \times 3.84 = 121$ nm, which is a substantial fraction of a wavelength.

However, the above scenario requires that the heated air blobs survive for long enough to fill most of the pipe. In fact, any deviations from temperature uniformity will decay away due to diffusion of heat from warmer regions to colder regions and the characteristic time constant for this decay will be small in a low-pressure environment. A sinusoidal temperature perturbation of wavelength λ decays with a time constant given by

$$\tau = \frac{1}{K}\left(\frac{\lambda}{2\pi}\right)^2,$$

where K is the thermal diffusivity given by

$$K = \frac{k}{C\rho},$$

with k being the conductivity, C being the specific heat capacity and ρ the density. The thermal diffusivity of air increases with decreasing air pressure because the conductivity remains roughly constant (providing the mean free path is smaller than the smallest dimensions being considered), and the specific heat remains roughly constant, but the density falls in proportion to the pressure. Substituting $\lambda = 40$ cm (i.e. a quasi-sinusoidal perturbation consisting of a 20-cm blob and a 20-cm region of unheated air) and a thermal difussivity appropriate for air at 1 mbar, we arrive at a time constant of around 200 ms. Thus, the heated air blobs are short-lived and unlikely to fill the pipe unless non-uniform sources of heating exist along the entire length of the pipe.

Polarisation

The reflections off mirrors in the beam train can induce changes in the state of polarisation of the beam on reflection. A quasi-monochromatic electro-magnetic wave can be modeled as the superposition of two perpendicularly polarised waves denoted S and P; by convention, the electric field of the S vibration is parallel to the line of intersection of the propagating wavefront with the surface of an optical component. The amplitude and phase reflection coefficients for mirror coatings depend on whether the light is S or P polarised. Typically the phase effects are more pronounced. For example, the amplitude reflection coefficient for light at a wavelength of 630 nm incident at an angle of incidence of 45° on a bare silver mirror is about 0.5% greater for S-polarized light than for P-polarized light, and the P-polarized light is phase-retarded by about 160° when compared with the S-polarized light. It is clear from this that the polarization effects of the optical train can be significant.

A general interference pattern can be thought of as the superposition of the interference patterns seen in two orthogonal polarisations, and if these two patterns have different fringe phases, the fringe contrast will be reduced. If the interferometer is constructed so that the polarisation effects of all the beam trains are identical, then the interference patterns in both polarisations will have the same phase and the fringe contrast will be maximised. This condition can be achieved if every light beam experiences the same set of polarisation changes in the same sequence as every other light beam. In practice, this means that the direction cosines between the direction of the beam and the mirror normal must be same for the nth reflection that the beam from telescope m experiences as for the nth reflection that any other beam experiences (Traub, 1988).

This path symmetry condition has been adopted in all the interferometers described in Section 4.1, and leads to a preference for certain geometries in the beam transport system. The NPOI, CHARA and MROI interferometers

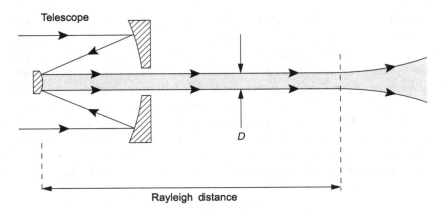

Figure 4.5 Diffraction of a beam propagating through the beam relay system.

have an equilateral 'Y'-shaped layout, so that the beams, after leaving the telescopes, experience reflections at 30° angles of incidence, while the beam transport within the VLTI has a rectangular geometry, with angles of incidence being generally 45°.

Diffraction

A circular parallel beam of light of diameter d and wavelength λ will propagate without significant diffraction effects over distances of up to

$$D_r = \pi d^2/4\lambda, \tag{4.1}$$

where D_r is known as the Rayleigh distance. At around this distance, Fresnel diffraction effects become important and the beam begins to spread, as shown in Figure 4.5. The eventual divergence angle of such a beam is given by the usual Fraunhofer diffraction angle $1.22\lambda/d$. For $\lambda = 1\mu m$, the Rayleigh distance for a 1-cm diameter beam is about 78 m, so diffraction effects are clearly important when propagating such beams over distances of hundreds of metres.

The effects of diffraction include both loss of light through spreading of the beam, and loss of fringe contrast because diffraction causes a mismatch in the phase and amplitude profiles of beams which have travelled different distances (Tango and Twiss, 1974). In order to keep these effects as small as possible, beam sizes of $d \gtrsim 10$ cm are typically used in optical interferometers. An alternative method of reducing the effects of diffraction is used in the VLTI, where a variable-curvature mirror is used to re-image the telescope pupil at the entrance to the beam-combining area (Ferrari *et al.*, 2000).

Diffraction occurs because of spatial non-uniformities in the wavefront. A finite beam diameter is a non-uniformity in the wavefront amplitude; wavefront distortions induced by atmospheric seeing can act as a non-uniformity in the wavefront phase. If the collectors are significantly larger than r_0 and have only partial adaptive correction, then the typical spatial scale of the seeing-induced phase non-uniformities will be smaller than that of the amplitude non-uniformity due to the beam size. Thus, the effective Rayleigh distance for these distortions will be less than for the aperture as a whole. Diffraction will cause the wavefront phase distortions to turn into amplitude distortions, and so a beam with an initially uniform intensity profile can take on a randomly changing 'speckled' appearance. If this beam then goes through an aperture which is smaller than the extent of the diffracted pattern, the total amount of light received will fluctuate randomly, which could cause problems in performing a photometric calibration of the system.

Although the effects of beam diffraction are mostly negative, they can be used to advantage: because any non-uniformities in the wavefront will tend to diffract out of the main beam, by choosing an appropriate size for the receiving aperture the wavefront distortions can be preferentially discarded while keeping the undistorted part of the beam; this is a form of spatial filtering (Horton et al., 2001), which does not require any additional optics.

4.4.3 Fibre beam transport

In radio interferometry the signal from the collectors is sent to the central correlator via wires or waveguides. The equivalent for optical interferometry is the use of optical fibres.

The fibres used for beam transport are single-mode optical fibres, for different reasons than those leading to their use in the context of spatial filtering as discussed in Section 3.7. Multi-mode fibres cannot be used for beam transport as different modes have different propagation speeds and so modes which are initially in phase will end up out of phase after propagation in the fibre. This effectively corrupts the optical wavefront and so can destroy the quality of an interference pattern.

Single-mode optical fibres have benefited from many years of development for telecommunication systems, and fibres with very low losses allow the transmission of optical signals over many kilometres. Compared to free-space propagation, fibres offer the possibility to simplify the 'plumbing' of interferometers, as fibres are far narrower and much more flexible than vacuum tubes acting as 'light pipes'.

Using fibres to get the light out from a moving telescope to a fixed beam-transport system can potentially save many reflections of the beam, and using fibres instead of systems of vacuum pipes between the telescopes and the beam-combining laboratory offers the possibility to reconfigure the array rapidly – literally a 'plug-and-play' system.

Fibres have been used successfully for beam transport over hundreds of metres in interferometric experiments (Perrin *et al.*, 2006b). However, none of the interferometers mentioned in Section 4.1 use fibres for transport of beams over distances of more than a few metres (for example within beam combiners – see Section 4.7.3), and there are a number of technical issues to be overcome before the use of fibres for beam transport becomes widespread.

One issue is to do with the wide waveband range typically used in astronomical interferometers. The transmission of silica optical fibres is very good at the near-infrared telecommunications wavelengths, but away from these wavelengths the transmission falls off.

To have a throughput of at least 50% over a 300-m length of fibre, the fibre needs to have losses lower than 0.01 dB/m, or equivelently 10 dB/km. Low-OH silica fibres can achieve this figure over a wavelength region from about 600 nm to 1900 nm but losses rise rapidly at shorter and longer wavelengths, so that the throughput of a 300-m length of fibre at a wavelength of 2200 nm is less than 5%. Fibres made from flouride and chalcogenide glasses can have improved performance in the near- and mid-infrared region, but have poorer performance at shorter wavelengths.

Another issue with the step-index fibres typically used in telecommunications is that they only operate efficiently as single-mode waveguides over a restricted range of wavelengths. At wavelengths shorter than the 'cut-off' wavelength characteristic for the fibre, the fibre becomes multi-mode, and at wavelengths more than about 1.35 times the cut-off, losses due to any bends of the fibres become prohibitive.

This problem has to some extent been overcome with *photonic-crystal* fibres, whose structure consists of a regular grid of air holes running along the length of the fibre (Birks *et al.*, 1997). These fibres can operate in single mode over an octave (i. e. factor of two) in wavelength, but they are a relatively recent development and have not yet been deployed in an interferometer.

As a result, to equip a broadband interferometer with a fibre-based beam transport system will likely require multiple fibres to accomodate the different bandpasses. This introduces layers of complexity not seen in more broadband optical systems. For example, it requires an optical system to split light at the different wavelengths at the telescopes, and means that a fringe tracker using the light in a bandpass going through one fibre will not take out any delay errors seen at wavelengths going through other fibres.

A second problem is that the intrinsic chromatic dispersion of the glass from which the fibres are made poses a problem when using the relatively wide bandwidths used to observe astronomical sources. This dispersion will cause a reduction in fringe contrast when the dispersion in the two arms of the interferometer being interfered is not equal. Nevertheless, a delay line introducing up to 2 m of differential delay has been built using fibres (Simohamed and Reynaud, 1997).

Finally, optical fibres are birefringent and so introduce large polarisation effects, which can reduce the fringe contrast by large factors. Static compensation of polarisation imbalances is possible for fibres in a thermally controlled environment but the birefringence depends on temperature and the level mechanical strain on the fibres caused by bends and twists, so dynamic control of the polarisation may be necessary if the fibres are outdoors.

4.5 Array layout

An array of M telescopes can simultaneously measure visibility information on $M(M - 1)/2$ different baselines. The question of how best to arrange these telescopes to achieve the best imaging performance has been the subject of many studies, particularly at radio wavelengths. These studies have resulted in the development of a number of 'rules of thumb' as to the type of (u, v) coverage which is generally to be preferred, although this to some extent depends on the target being observed:

1. If a target has a certain overall size θ_{max} and contains details of interest separated by θ_{min} then the arguments in Section 2.3 suggest that the (u, v) plane needs to be sampled at a spacing of about $\Delta u \sim 1/\theta_{max}$ up to a maximum (u, v) coordinate of magnitude $1/\theta_{min}$.
2. Multiple (u, v)-plane samples spaced by much less than $1/\theta_{max}$ are less useful than the equivalent number of samples positioned to 'fill in' less-well-sampled areas of the (u, v) plane. Samples spaced by $1/\theta_{max}$ or more are termed 'independent (u, v) samples' whereas arrays that sample the same (u, v) coordinate multiple times are termed 'redundant arrays'.
3. The above rules can be relaxed for targets which are known to be 'sparse' in the sense that an image of the target at the desired resolution has relatively few resolution elements which contain flux. For example, a binary

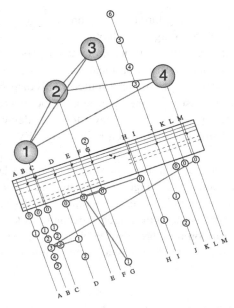

Figure 4.6 Array layout for the VLTI. The layout includes both the 8-m-diameter unit telescopes (shown as large circles) and the 30 pads for the 1.8-m-diameter auxiliary telescopes (shown as smaller circles). The baselines that were used for calibrator observations by Richichi and Percheron (2005) are overlaid on the diagram.

star system imaged at a resolution which allows the two stars to be distinguished but does not significantly resolve the stars themselves will have two 'filled pixels' in the image. In the case of a sparse target, image reconstruction can usually be accomplished when the number of independent (u, v) points sampled approximately equals or exceeds the number of filled pixels.

Despite the large number of studies, no single arrangement of telescopes has been found to be best in all cases and there are a large number of geometries to choose from which give adequate (u, v) coverage. The final choice of telescope layout tends to be made based on other considerations, for example the beam transport system symmetry concerns outlined in Section 4.4 or fitting into the local topography. Two such layouts are shown in Figure 4.6 and Figure 4.7, which are based on two different beam-relay geometries, one rectangular and one 'Y'-shaped.

If the telescopes in the array are relocatable, then the (u, v) coverage can be increased by rearranging the telescopes into a number of different arrays.

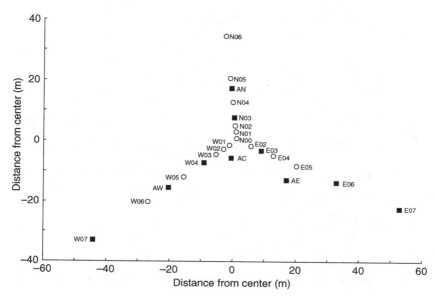

Figure 4.7 Array layout for the inner portion of the NPOI. The siderostat stations active in the late-2014 observing season are shown as filled squares. The fixed 'astrometric' stations have labels beginning with 'A', while the 'imaging' stations, which can take movable siderostats, have labels beginning with 'N', 'E' or 'W', depending on which arm of the array they are on. The outer imaging stations of the array are not shown: these extend to baselines of up to 450 m (image courtesy Don Hutter).

This is done for example with the auxilliary telescopes (ATs) on the VLTI, where there are 30 different stations to place the four ATs on. A number of different 'quadruplets' of AT stations are offered in any given semester. Arrays such as the NPOI and MROI offer multiple array configurations, which are approximately scaled versions of each other. These allow a given level of imaging performance to be matched to similarly-shaped targets of different angular sizes.

4.6 Delay lines

Delay lines are also called 'path compensators' to indicate that their function is to compensate for any imbalance in the delays introduced elsewhere in the paths from the target to the beam combiner. The amount of differential delay that needs to be introduced can be separated into a static component and a dynamic component. The static component is typically dominated by the differential paths introduced by the beam transport optics. The dynamic

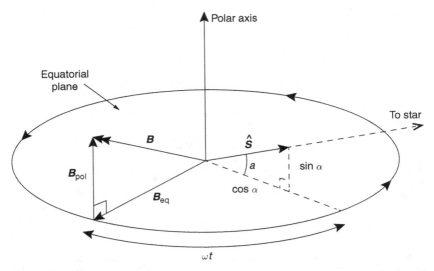

Figure 4.8 Geometry for calculating the change of delay with time due to Earth rotation.

component is dominated by the variation in the geometric delay when observing different parts of the sky, either when observing different targets or when observing a single target as the Earth rotates, but it also includes smaller components due to drifts and vibrations in the instrument and to atmospheric seeing.

In Section 1.4.5 it was shown that the geometric delay for light from a target in a direction $\hat{S}$ and a pair of telescopes i and j with separation $\boldsymbol{B}_{ij}$ is $d_{\text{geometric}} = \boldsymbol{B}_{ij} \cdot \hat{S}$. Thus, the maximum and minimum delays occur when the target lies in directions parallel or anti-parallel to the baseline, and so the delay ranges from $-|B_{ij}|$ to $|B_{ij}|$. In practice targets in these directions are too close to the horizon to observe; if targets are only observed if they are at least 30° above the horizon, then this range is reduced to $\pm \sqrt{3}|B_{ij}|/2$. Thus the delay range required increases linearly with the baseline and can easily reach hundreds of metres.

As explained in Section 2.3, any Earth-based baseline rotates with respect to a celestial target. This causes the required delay to change with time even when observing a single target. The maximum rate of change of delay can be deduced by decomposing the baseline vector $\boldsymbol{B}$ into a fixed component parallel to the polar axis of the Earth $\boldsymbol{B}_{\text{pol}}$ and a component in the equatorial plane $\boldsymbol{B}_{\text{eq}}$, which rotates as shown in Figure 4.8. The geometric delay is then given by

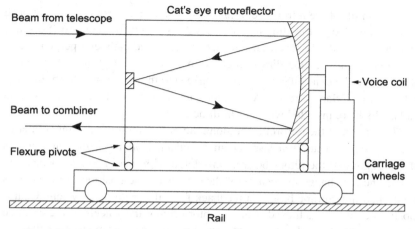

Figure 4.9 Schematic diagram of a delay line showing the path of the starlight through the cat's eye retroreflector. The path of the laser metrology beam has the same geometry as for the starlight beam, but in a plane perpendicular to the page.

$$d_{\text{geometric}} = |\boldsymbol{B}_{\text{pol}}| \sin \alpha + |\boldsymbol{B}_{\text{eq}}| \cos \alpha \cos \omega t, \qquad (4.2)$$

where α is the declination of the object, ω is the rate of rotation of the Earth and t is zero at the moment that the angle between the projections of $\hat{\boldsymbol{S}}$ and $\boldsymbol{B}$ onto the equatorial plane is zero. The rate of change of delay is therefore

$$\dot{d}_{\text{geometric}} = \omega |\boldsymbol{B}_{\text{eq}}| \cos \alpha \sin \omega t, \qquad (4.3)$$

which has a maximum/minimum value of $\pm \omega |\boldsymbol{B}_{\text{eq}}|$ when the object is at zero declination. Thus, for $|\boldsymbol{B}_{\text{eq}}| = 100\,\text{m}$, the rate of change of delay reaches a maximum value of 7 mm/s. This change of delay is required to happen smoothly: the short-term 'jitter' in the delay due to vibrations etc. in each delay line has to be kept below $\lambda/20$ during an exposure in order to keep the visibility losses to less than 10%, which translates into a jitter of less than 15 nm RMS at a visible wavelength of 600 nm.

The requirement for a factor of order 10^{10} between the delay range and the allowable delay jitter is met using the multi-stage servo concept first developed by Pierre Connes (Connes and Michel, 1975). An example layout for such a system is shown in Figure 4.9. Starlight is reflected off a retroreflector, which is mounted via a flexible suspension system to a carriage that can be moved to adjust the distance travelled by the beams.

The retroreflector shown in the figure is a so-called 'cat's eye' arrangement consisting of a paraboloid primary mirror with a flat secondary mirror at its focus. This has the advantage that the tilt of the reflected beam is insensitive to changes in tilt of the retroreflector and so the tilt of the carriage does not need to be controlled at the sub-arcsecond level, as would be required if a plane mirror was used. However, any displacement of the cat's eye perpendicular to the beam propagation direction causes a displacement of the output beam; this needs to be controlled in order to make sure that the beams from different telescopes overlap correctly. As a result, the carriage is often made to run on rails, which are installed with sub-millimetre precision.

The motion of the retroreflector along the rails is monitored using a 'metrology system' – usually a laser beam is reflected off the same cat's eye as the starlight and the return beam is interfered with a reference laser beam to determine the change in position of the cat's eye. The position of the latter is controlled in real time based on feedback from the metrology system. Coarse position control is achieved using a motor driving the wheels of the carriage and fine control is achieved by moving the retroreflector relative to the carriage using a position actuator, typically an electromagnetic actuator (known as a 'voice coil') or a piezoelectric actuator. Using this technique, delays of hundreds of metres can be introduced with an error measured in nanometres.

Running the delay lines inside a vacuum system eliminates the chromatic dispersion described in Section 1.8 and becomes mandatory for delays of more than 100 m or so. Installing polished rails at high precision inside a vacuum system can become expensive as the length increases; therefore, the CHARA array and the NPOI use a pair of delay lines in series, a short length of continuously variable delay and a longer delay line, which can introduce discrete static delays. Because the longer delay line is static during an observation, it can be based on technologies which do not require precision rails such as 'pop-up' mirrors at discrete locations.

An alternative is to construct a continuously variable delay line, which does not require precision rails. At the MROI, the carriage runs directly on the inside of standard pipes acting as the vacuum enclosure (Fisher *et al.*, 2010). A control system monitors and corrects any errors in the transverse location ('shear') of the reflected beams caused by imperfections in the pipe manufacture. As a result, up to 400 m of delay can be introduced in a single stage. This arrangement allows delays to be changed more rapidly and has higher optical throughput than a two-stage system with switched mirrors.

4.7 Beam combiners

The beams from the different telescopes are mixed to form interference fringes in the beam combiner. There are two classes of information provided by beam

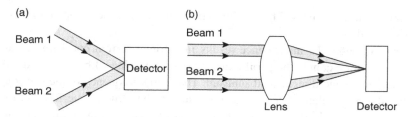

Figure 4.10 Pupil-plane (a) and image-plane (b) beam combination.

combiners. The first is information on the amplitude and phase of the object visibility function, and the second is real-time information on the piston disturbances due to atmospheric seeing and instrumental vibrations and drifts. Both types of information can be produced by a single beam combiner, but the discussion here will centre on the idea of using two specialised combiners, each producing one kind of information, called the 'science combiner' and the 'fringe-tracking combiner', respectively. The science beam combiner will be the main focus of the current section; fringe-tracking combiners will be discussed briefly at the end of this section.

There are a remarkably diverse set of ways of achieving the function of beam combination and so a text like this cannot hope to describe all the possibilities in depth. Instead, the 'zoo' of possible beam combiners is described here in terms of its composition out of a set of functional elements, followed by a few practical examples.

4.7.1 Free-space combination

A major subdivision of types of beam combiners is into free-space and guided-optics combiners. As might be expected from the name, in free-space beam combiners the beam combination takes place under conditions where the beams are propagating without confinement, while in guided-optics combiners, which are discussed in Section 4.7.5, the beams are combined while propagating in waveguides whose dimensions are of order a wavelength.

One element of a free-space beam-combiner design is whether it makes use of 'pupil-plane' or 'image-plane' interference as shown in Figure 4.10. A pupil-plane design was used in the example interferometer introduced in Chapter 1.3 and simply superposes images (usually demagnified) of the telescope pupils. In an image-plane design the light from multiple pupils is brought to a focus in a plane where images of the target as seen through the telescopes (for point-like targets this will be the diffraction pattern of the telescope pupil) are superposed.

In both cases the interference results in a fringe pattern consisting of an envelope (for example the pupil image or the image-plane diffraction pattern) modulating a one-dimensional sinusoid. The spatial frequency of the sinusoid depends on the relative tilt between the beams at the plane of interference; for the image-plane design, this tilt is proportional to the spacing of the beams in the 'input pupil' of the focussing optic.

A potential advantage of the pupil-plane scheme is that tip–tilt errors introduced by atmospheric seeing change the fringe frequency but do not reduce the fringe contrast: in the case of image-plane interference, tip–tilt errors shift the images and the reduction in image overlap causes a reduction in fringe contrast.

In practice, this advantage of pupil-plane interferometry is limited because the SNR of the fringes is seldom high enough to determine the change in fringe frequency on a frame-by-frame basis. If the power is measured at a fixed frequency in the fringe pattern then atmospheric tilts cause a loss in fringe power comparable to that seen in the equivalent image-plane combiner.

A unique advantage of image-plane combiners occurs when they are operated in a so-called "Fizeau" mode (Faucherre *et al.*, 1990). In this mode the beams are arranged in the combiner as a scaled version of the input pupil of the interferometer as a whole, that is to say the wavefront sampled by the telescopes. This means that the vector separations of the beams at the focussing optic must be a scaled version of the interferometric vector baseline, with a demagnification the same as has occurred to the individual beams.

If this condition is satisfied, then fringes can be seen over a field of view which is considerably larger than the diffraction limit of the individual telescopes. This is because the optical path difference (OPD) introduced between the interfering beams an object not being at the phase centre is exactly cancelled by the change in OPD induced for the equivalent point in the image plane where the beams meet.

In practice, the field of view of an interferometer is usually limited to values considerably less than the diffraction limit of the collectors by the limited sampling of the (u, v) plane (see Section 2.4.3), and so no Fizeau-mode combiners have so far been implemented in long-baseline interferometers: all the Fizeau-mode combiners in use are in aperture-masking systems or in systems where the collectors are on a common mount such as the LBT.

One form of the pupil-plane combiner makes use of a beamsplitter to combine the pupil images as shown in Figure 4.11. In this configuration, the relative tilt between the combined beams can be made to be zero; in this 'co-axial' case

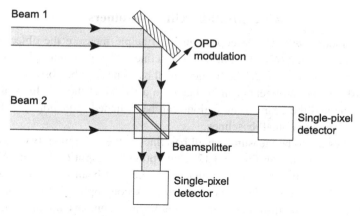

Figure 4.11 Co-axial temporally coded fringe pattern.

the spatial frequency of the fringes tends towards zero, in other words each beam is uniform in intensity and the fringes are said to be 'fluffed out'. Each output can be sampled using a single-pixel detector and the measured intensity at any moment depends on the phase difference between the beams: if the path length in one arm of the interferometer is changed linearly with time, the intensity will vary sinusoidally with time – a 'temporal fringe pattern', which can be captured by recording the detector output as a function of time. Thus, we can speak of fringes being 'spatially coded' or 'temporally coded', and it is also possible to have a mixture of the two.

Typically, the differential delay is scanned backwards and forwards rapidly enough to capture the fringe information before the fringe phase is altered by atmospheric seeing. The delay scanning (or delay *modulation*) is usually accomplished using mirrors attached to fast electro-mechanical actuators such as piezoelectric devices or 'voice-coil' actuators, although it is in principle possible to use electro-optic devices such as liquid crystals to perform the same function.

Note that this arrangement has two optical outputs: energy conservation means that when one output is bright, the other is dark (assuming a lossless beamsplitter). Thus, the outputs are said to be *complementary* and the temporal fringe patterns seen in the two outputs are 180° out of phase; subtracting the intensities of the two outputs gives an estimate for the sinusoidal fringe signal independent of any fluctuations in the total signal level, and so is useful in situations where there is a slowly-fluctuating background, for example in the thermal infrared.

4.7.2 Multi-baseline combiners

Interferometers with N telescopes can in principle measure the object visibility function on $N(N-1)/2$ different baselines. There are multiple ways to accomplish this for $N > 2$. The fringes on all baselines can be observed simultaneously or measurements can be made on a subset of the baselines at any given time and the beam combinations can be switched sequentially to build up the information on all baselines.

If fringes are to be measured on all baselines at the same time, two ways of doing this are shown in Figure 4.12. One option is to split each beam $N-1$ ways, and perform pairwise combination of the split beams to form a set of $N(N-1)/2$ two-beam interference patterns. A second option is to combine up to N beams into an 'all-in-one' fringe pattern, which contains the superposition of $N(N-1)/2$ fringe patterns corresponding to interference between all pairs of beams. Other options exist, for example combining beams in M-way fringe patterns, where $2 < M < N$.

For all-in-one combination, the superposed fringe patterns need to be 'multiplexed' (or 'coded') in some way, which allows the information corresponding to the visibilities on different baselines to be separated from each other, in other words 'demultiplexed' (or 'decoded'). The most common way of achieving this multiplexing is to arrange for the different fringe patterns to have different fringe frequencies.

Figure 4.13 shows a three-beam image-plane combiner where the input beams are arranged in a line with nearest-neighbour centre-to-centre spacings of 1:2. The fringe frequency for the interference between a given pair of beams will be proportional to the spacing between the beams, and so the fringe frequencies for the three possible baselines will be in the ratio 1:2:3. Fringe information for each baseline can be extracted by taking the Fourier components of the multiplexed fringe pattern at each of the three spatial frequencies as shown in Figure 4.13.

The arrangement of the input beams in this example is a 'non-redundant' one, in that none of the three spacings between the beams overlaps with any other spacing. Such non-redundant spacings are equally useful in choosing the best spacings for telescopes in an interferometric array, and so are a well-researched subject. Two-dimensional non-redundant arrays can also be used as well as the one-dimensional array shown in this example, but are less useful when spectral dispersion is incorporated, as described in Section 4.7.4.

A non-redundant pattern can be used similarly for a co-axial temporally multiplexed system as shown in Figure 4.14. In this case the fringe frequencies are proportional to the difference between the delay scanning *velocities* for

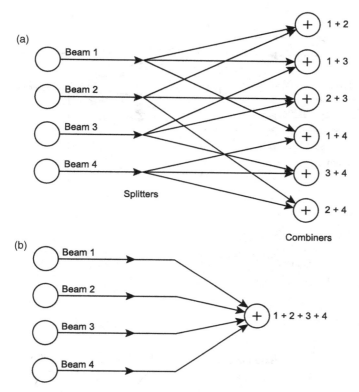

Figure 4.12 Schematic outline of two types of multi-telescope beam combiners, a pairwise combiner (a) and an all-in-one combiner (b).

the relevant pair of beams, and so the velocities must be arranged in a non-redundant fashion as shown in Figure 4.15.

An issue for multi-baseline combiners is *crosstalk* between the fringes on different baselines: if some of the fringe signal from one baseline leaks into the signal measurement for another baseline this can introduce errors in the visibility measurement, which are particularly difficult to calibrate as they will depend on the visibilities on different baselines. Crosstalk can arise if the fringe frequencies of different baselines are not sufficiently well separated or can be caused by atmospheric temporal piston variations during the scan in temporally multiplexed interferometers (Buscher, 1988b).

4.7.3 Spatial filtering

The use of spatial filtering to improve fringe measurement in the presence of atmospheric seeing has been described in Section 3.7. Spatial filtering is

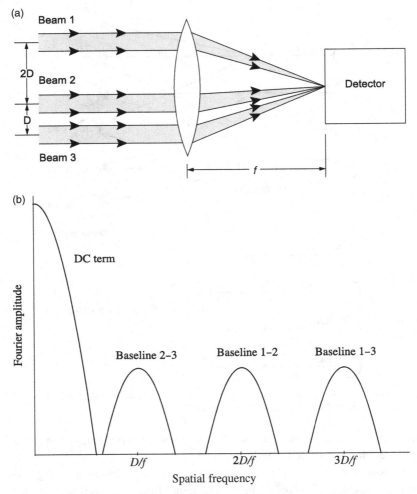

Figure 4.13 Multiplexing three beams into a single fringe pattern in the image plane (a) and the modulus of the Fourier transform of the fringe pattern showing peaks at three different frequencies (b).

usually accomplished within the beam combiner for a number of practical reasons.

In an image-plane combiner, a form of spatial filtering can be accomplished in software simply by setting to zero (or not reading) the intensity of any pixels outside a 'virtual pinhole' of a chosen size – usually the diameter is chosen to be comparable to the diffraction limit of the telescope. This is entirely

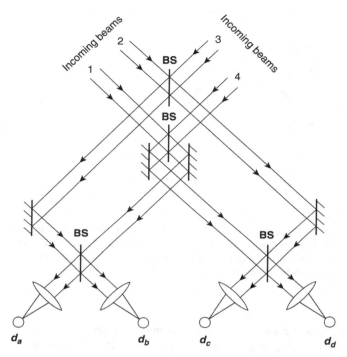

Figure 4.14 Schematic for the four-way temporally multiplexed beam combiner
in the COAST interferometer (Baldwin *et al.*, 1996). BS indicates beamsplitters
where the beams are combined. The output is on four detectors d_a–d_d, which all
receive the superposition of all four input beams. OPD modulation takes place
external to the beam combiner.

equivalent to passing each of the beams through a real pinhole of the same size
before combination, but requires no additional hardware.

In a co-axial beam combiner, spatial filtering can be straightforwardly
accomplished by focussing the combined light after beam combination onto
a single-mode fibre. Since all the light is filtered by the same fibre, there is no
need to match the lengths of the fibres to equalise chromatic dispersion and
polarisation effects, as is needed when separate fibres are used for each input
beam.

Single-mode fibres at the input of a beam combiner can be used to perform three functions simultaneously: spatial filtering, beam transport and beam
reconfiguration. For example, in the AMBER beam combiner, the light from
each of three telescopes illuminates a separate single-mode fibre. The fibres
transport the light a few metres from the room-temperature environment into

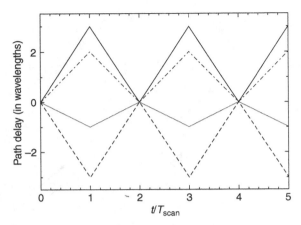

Figure 4.15 The scanning pattern for the temporally multiplexed combiner in Figure 4.14. The scanning velocities are at −3, −1, +2, and +3 times a base velocity. This non-redundant arrangement of velocity differences means that the interference fringes between the six pairs of input beams all appear at different fringe frequencies.

a cryogenic dewar and bring the beams into an arrangement where they are all arranged in a linear non-redundant pattern to illuminate an 'image-plane' combiner as shown in Figure 4.16 (the distinction between the image plane and the pupil plane is somewhat arbitrary after the beams have been through a single-mode fibre, as the beam intensity profile in all planes is roughly Gaussian).

4.7.4 Spectro-interferometry

Interference is a wavelength-dependent phenomenon, so there are a number of practical reasons, for example minimising the effects of chromatic dispersion, for observing interference in narrow spectral bands. In addition, much of the physical information about an astronomical object is derived from looking at the variation with wavelength of the emission from the object (or part of the object) under study. Thus, the combination of spectral information and interferometry can potentially provide better interferometric performance and more information about the object.

The simplest form of spectro-interferometry can be achieved by placing a narrow-band filter in front of the detector, observing the interference fringes, and then repeating the observation with a different filter. Spectrally dispersing the light by placing a prism or grating in the beam-combination optics allows

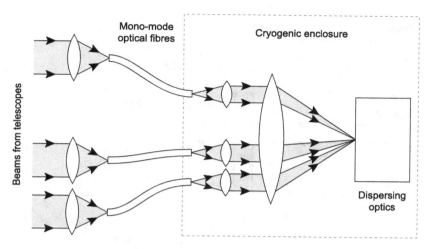

Figure 4.16 Schematic of the AMBER beam combiner.

interference to be observed simultaneously at many wavelengths and so spectro-interferometry of this form is now used in most optical interferometers.

Spectro-interferometry can be simply implemented in a co-axial temporally coded fringe pattern by placing a prism in the combined beam and focussing the resulting spectrum onto a one-dimensional detector as shown in Figure 4.17. Each pixel in the detector will show temporally coded fringes and can be analysed as a parallel collection of interferometers working in different bandpasses.

Spectral dispersion typically spreads the signal continuously along one dimension of a two-dimensional detector; therefore, in a spatially coded beam combiner the fringe modulation must be arranged to be in the perpendicular direction to this, in order that the fringe information is not 'smeared out'. Thus, the fringes for different baselines must run parallel to each other and so one dimension of the monochromatic fringe pattern contains no information. Most combiner designs will 'squash' the monochromatic fringe pattern in this dimension using anamorphic optics in order to reduce the number of pixels needed to sample the pattern, and then disperse the light along the direction in which the fringe pattern has been squashed to give a dispersed fringe pattern as shown in Figure 4.18.

An alternative form of spectro-interferometry combines two separate interferometric techniques: spatial interferometry and Fourier-transform spectroscopy. Such 'double-Fourier' interferometry (Mariotti and Ridgway, 1988) can be understood in the context of a co-axial, temporally coded beam

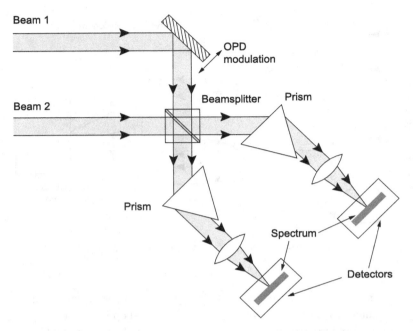

Figure 4.17 A spectrally dispersed temporally coded beam combiner.

Figure 4.18 Spectrally dispersed fringes from the AMBER beam combiner. The spectral dispersion is in the vertical direction: each horizontal line on the right-hand side is a spatially coded fringe at a given wavelength. The left-hand pair of channels are photometric calibration channels and so have no fringes. From Tatulli and Duvert (2007).

combiner with no spectral dispersion, illuminated by polychromatic radiation. As explained in Section 1.7, the fringe pattern seen with wideband radiation depends on a combination of the object visibility function and a Fourier transform of the intensity spectrum of the source. This combination can be recovered by observing the fringe pattern over a wide range of OPD, and performing the Fourier transform of the intensity as a function of OPD. In other words, the spectral dispersion function is accomplished using the Fourier-transform spectrometer that is built in to any interferometer, rather than needing a separate spectral dispersing element.

4.7.5 Guided-optics combination

Beam combiners using lenses or beamsplitters often have strict alignment requirements in order to maintain high fringe contrast and as a result can require frequent and sometimes laborious realignment. One way around this is to make use of guided optics to keep the light confined right up to the point of beam combination.

One form of guided-optics combination uses single-mode optical-fibre couplers instead of beamsplitters. This form of beam combination is represented by the FLUOR (Coudé du Foresto *et al.*, 2003) and VINCI (Ségransan *et al.*, 2003) combiners on the CHARA array and the VLTI, respectively. Propagation in single-mode fibres provides spatial filtering and these combiners have secondary outputs for photometric calibration (see Section 8.8.2), and so both combiners are able to provide high-precision visibility measurements.

Both the FLUOR and VINCI combiners are two-telescope combiners. Multi-way combiners using fibre couplers are possible in principle but become complex. A more scalable technology is integrated optics, the optical equivalent to integrated circuits. Complete optical systems can be fabricated using microlithography techniques, and because all the optical components are built into a single 'chip' a few millimetres in size, then once the device is fabricated there is little opportunity for components to go out of alignment and the effects of thermal expansion are usually small.

Integrated optics are typically fabricated out of planar optical waveguides fabricated in a two-dimensional 'wafer'. Because the transverse dimensions of the waveguide are smaller than a wavelength, they act as monomode devices, similar to optical fibres, so provide spatial filtering to the beams.

An example of an integrated optics combiner is the PIONIER combiner used on the VLTI (Le Bouquin *et al.*, 2011). The layout of the combiner is shown in Figure 4.19. Light from one of the VLTI telescopes is injected into each of the four inputs to the combiners. The 'optical circuit' on the integrated optics chip

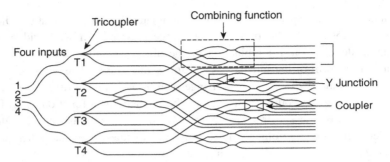

Figure 4.19 The layout of the integrated optics chip used in the PIONIER beam combiner. From Benisty *et al.* (2009).

implements a set of pairwise co-axial beam combiners to make interference for all six baselines simultaneously. In fact there are two beam combiners for each pair of telescopes: one combiner produces fringes in its complementary outputs at phase shifts (with respect to some arbitrary zero point of phase) of 0° and 180° and the second combiner produces fringes with phase shifts of approximately 90° and 270°. This so-called quadrature or 'ABCD' fringe information is sufficient to measure the amplitude and phase of the fringes (see Section 8.5.3) without requiring additional temporal fringe scanning.

4.7.6 Fringe-tracking combiners

Fringe-tracking beam combiners have similar requirements to science combiners, but the weighting between these requirements is different. Although in both cases the combiner must be designed to ensure accuracy in the measurements of fringe phase, in a fringe tracker the measurement of fringe amplitude is less important. In addition, the desirability of sampling the fringe visibility on all $N(N-1)$ baselines in an N-telescope interferometer is reduced in the case of a fringe tracker, since all the atmospheric OPD values can be sampled using $N-1$ appropriately chosen baselines. Finally, the scientific requirement for a large amount of spectral resolving power in order, for example, to isolate specific spectral features in the target is not present.

Instead, a primary requirement for the fringe tracker is to achieve as high an SNR on the fringe phase as possible in as short a time as possible in order to follow the atmospheric perturbations. Optimising for this requirement might lead to a design which has, for example, lower spectral resolution (or none at all) and covers fewer baselines than an equivalent science combiner.

An example fringe-tracking combiner is the CHAMP (Berger *et al.*, 2008; Monnier *et al.*, 2012) combiner at the CHARA array. Dichroic (i.e. wavelength-dependent) beamsplitters take light from all six of the CHARA telescopes, but instead of measuring fringes on all 15 baselines, the combiner performs a set of six pairwise fringe combinations.

The beam combination is a co-axial, temporally modulated design. Each output of the combiner is focussed onto a single pixel of an infrared array – there is no spectral dispersion: the light from a whole astronomical band (J, H or K band) is used. The design is optimised for sensitivity at low light levels when using readout-noise-dominated detectors: the fewest pixel readouts are done in order to maximise the SNR in a single read.

4.8 Detectors

The beam combiner forms a fringe pattern on one or more light-sensitive devices. These 'detectors' are responsible for converting the light intensity pattern into a form suitable for analysis. The detector used in Michelson's day was the human eye, but modern interferometers use photoelectric devices, which convert the intensity into an electronic signal. These detectors are far more sensitive than the human eye, but nevertheless have limitations that can be critical to the performance of the interferometer.

Typical detectors are array devices, consisting of a two-dimensional grid of light-sensitive elements or 'pixels'. The light level in all the pixels are typically read out as a 'frame' at the end of each exposure. An important requirement for interferometric detectors is the ability to operate at frame rates of hundreds of times per second in order to avoid temporal smearing of the fringes. This is quite different from most detectors used in astronomical applications, which are typically read out less than one frame per second. The frame rate can often be increased by reading out fewer pixels; this requires efficient beam-combiner designs which put all the light into as few pixels as possible.

A second consequence of this high frame rate is that there will be relatively few photons captured per frame, so maximising the amount of light captured (known as the 'quantum efficiency' or QE) and minimising the amount of readout noise (see Section 5.2.2) is critical for interferometric detectors.

At optical wavelengths there are two types of high-frame-rate device with negligible readout noise and good QE. Silicon avalanche photodiodes (APDs) are photodiodes with internal electrical gain. When operated in so-called 'Geiger mode' the effective gain can be more than 10^6 so that they produce a macroscopically detectable pulse of current for the arrival of each photon. Photon-counting techniques mean that these devices can be entirely photon-noise-limited and the QE can be as high as 60% or more. A disadvantage of APDs is that they are usually only single-pixel devices (or effectively so – the trend is towards multiple diodes, which are wired together as a single pixel) and so designs that use only a few pixels, such as co-axial combiners, are favoured when using silicon APDs.

A newer technology is electron-multiplying CCDs, in which the light detection is in a two-dimensional silicon array of thousands or millions of pixels, but which include amplification similar to that of APDs. These can achieve photon-counting performance under low-light-level conditions and can have quantum efficiencies of more than 90%; they are therefore becoming the detector of choice for interferometric applications.

Until recently the situation at near-infrared wavelengths was not as good as in the optical, because the read noise of high-QE detectors for this wavelength regime was at the level of 10 electrons per read or more. Furthermore, to keep the readout noise low, the readout had to be relatively slow so that designs which involved large numbers of pixel reads were less favoured.

This has begun to change because of the advent of detectors which include APDs made of HgCdTe, a semiconductor which has good QE (typically above 60%) at infrared wavelengths. Devices having read noises as low as 0.8 electrons RMS have been tested in the laboratory, so that it is looking likely that photon counting at these wavelengths may become possible in the near future (Finger *et al.*, 2012). Furthermore, these APDs can be straightforwardly incorporated into two-dimensional arrays and it is therefore likely that this technology will come to dominate near-infrared interferometry.

At wavelengths longwards of 3 μm the dominant source of noise is likely to be thermal background noise and so read noise is less of an issue. The main detector characteristics that are important under background-limited conditions are linearity and uniformity of the response between different pixels, so that the large background levels can be accurately subtracted.

4.9 Alignment

Alignment of the optical components in the beam train is a serious practical issue in all interferometers, because the mechanical tolerances to which the optical elements need to be placed are severe. The approach used in all interferometers is not to try to place components by 'dead reckoning' alone, but rather to adjust the positions of the components based on optical measurements of the error in position.

An interferometer needs to be aligned when it is initially assembled, but the componenents will inevitably drift out of alignment due to thermal and other long-term mechanical effects. The timescales before these drifts become noticeable can be as long as months or as short as minutes but inevitably realignment is necessary, and so an ongoing programme of realignment is part of the operation of any interferometer.

4.9.1 Piston

The typical alignment needs of an interferometer can be broken down in terms of the Zernike components of the wavefront error affected. The lowest-order component, the piston component, is the one subject to the most severe drifts. Thermally induced expansion and contraction effects can mean that the pathlengths inside an interferometer and the baselines within a telescope can change by hundreds of microns over the course of a day, much greater than the RMS errors introduced by the atmosphere and typically larger than the coherence lengths over which fringes can be found.

Piston errors are typically compensated for by adjusting the delays introduced by the delay lines. In the crudest case, the delay lines are scanned backwards and forwards until fringes are found on each baseline on each star.

A more sophisticated approach makes use of a model which tries to predict where the fringes can be found, so that the time spent searching for fringes is minimised. This 'baseline model' typically includes parameters describing the three-dimensional locations of the telescopes and the internal delays from the telescopes to the beam combiner. This latter parameter is sometimes called the 'constant term' as the locations of the telescopes predict the external OPD due to the location of the star in the sky, which is changing, while the internal OPD is stable at least over the short term.

This model is updated using measurements of the delay offsets at which fringes are found on stars; after the measurement of the fringe delay offsets on at least four stars widely dispersed across the sky the whole model can be updated.

4.9.2 Tilt

The tip–tilt Zernike aberrations typically experience drifts, which are not as severe as those for piston, but are still at the level where regular readjustment is mandatory.

Any errors in the tilts of any mirror in the system will cause twice that error in the final tilt of the beams at the beam combiner. For a differential tilt error between two circular beams of θ radians the fringe contrast will be reduced by a factor

$$\gamma \approx 1 - (\pi\theta D/\lambda)^2/8, \tag{4.4}$$

where D is the diametre of the beam (Porro et al., 1999).

With a beam diameter of 10 cm, a drift in the tilt of one mirror of about 0.05 arcsecond will cause a 1% degradation of fringe contrast for light at a wavelength of 500 nm. This order of magnitude of tilt stability is difficult to attain so some kind of tilt realignment is usually necessary on at least a nightly basis.

Realignment of the net tilt of the optical train can be accomplished by using a tip–tilt sensor close to the beam combiner, i. e. after most of the optical elements in the beam train. A complication is that full realignment requires more information than is available from beam tilt measurements made at one location in the beam train: because of the long pathlengths inside an interferometer, tilt drifts can cause the transverse position of the beam (the so called 'beam shear') to drift as well.

For example, a 1-arcsecond tilt error in a mirror 500 m from the beam combiner can cause the beam centre to move by 5 mm, enough to cause a significant loss of beam overlap at the beam combiner and hence a loss in fringe contrast. Larger beam-shear drifts can cause the beam to 'wander off' the mirrors in the beam train and so light will be lost due to vignetting. Mirrors which are tilted by the same amount but are different distances from the combiner will cause different amounts of beam shear, so without knowing which mirrors have drifted, it is difficult to know what combination of tilt and shear correction is required.

The solution usually adopted is to make measurements at multiple places along the beam train to constain the tilt errors of most or all of the mirrors in the train. There are perhaps as many ways of effecting this alignment as there are interferometers, but a common feature is that alignment of every element in the beam train is typically time-consuming and so is not performed very often. As a result, significant errors in beam tilt and shear can build up as a result of instrumental drifts between system realignments.

An approach to remedy this is to combine relatively infrequent 'coarse' alignment to correct any gross errors in individual optics, which could cause vignetting, and more frequent 'fine' realignment of a subset of the optics to correct for the drifts in the overall tilt and shear of the beams, which could cause reductions in fringe contrast. If the realignment procedure is automated, it can be repeated frequently without too much loss in observing time.

4.9.3 Higher-order errors

Higher-order wavefront errors such as defocus are less of a problem, as they are less sensitive to mechanical alignment issues. For example, in most systems wavefront defocus can be held at an adequate level by tolerancing of spacings

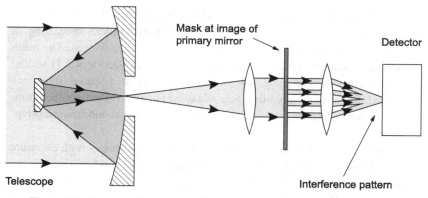

Figure 4.20 The optical setup for an aperture-masking experiment on a large single telescope.

at the micron level, while tilt errors can be significant when the mechanical errors reach a few tens of nanometres. As a result many of the higher-order errors can be removed at the initial alignment stage and no further realignment is necessary – 'set and forget'. An exception to this is the telescopes, which are in a thermally hostile environment and typically need to be focussed on at least a nightly basis.

4.10 Aperture masking

Interferometry with a single telescope sounds like a contradiction in terms, but at optical and infrared wavelengths this technique is currently one of the most scientifically productive ways of imaging using interferometry. The first interferometric measurement of an astronomical object was done this way (Stephan, 1874), and the technique was revived when closure phase was shown to be a practicable way of imaging at optical wavelengths in the presence of turbulence (Haniff et al., 1987).

Aperture masking works by converting a single large aperture into multiple smaller apertures using a mask containing an array of small holes of order r_0 in diameter. This mask is nominally located at the entrance aperture of the telescope, but for practical convenience is usually mounted at a reimaged and demagnified pupil plane in an optical arrangement like that shown in Figure 4.20, or in some cases at the secondary mirror of the telescope (Tuthill et al., 2000b).

The light passing through the holes comes together in the focal plane of the telescope and forms an image-plane interference pattern consisting of the diffraction pattern of the holes crossed by interference fringes. The holes are normally arranged in a non-redundant pattern (see Section 4.7.1) so that the interference between any pair of holes appears at a spatial frequency which is different from any other pair of holes. For this reason another common term for aperture-masking interferometry is 'non-redundant masking' or NRM.

A camera mounted in the focal plane is used to take images with exposure times of order t_0 in order to 'freeze' the fringe motions and the images are processed in an identical manner to the fringe patterns from a long-baseline interferometer. If data at sufficient (u, v) points are collected, images can be reconstructed from the power spectrum and bispectrum and will have a maximum resolution of order λ/D, where D, is the diameter of the telescope.

This may not seem any better than can be achieved with adaptive optics on the same telescope, but there are a number advantages over AO which make NRM scientifically competitive. The first of these is that NRM can work well at shorter wavelengths where AO struggles to perform.

Secondly, even at the near-infrared wavelengths that modern AO systems are optimised for, AO systems typically leave behind some uncorrected wavefront perturbations. Propagating these perturbations to the image plane results in an image of a point object which consists of a diffraction-limited 'core' surrounded by a more diffuse 'halo'.

If there is faint structure near to a bright object this can be lost in the halo, particularly as this halo is time-variable and therefore is difficult to calibrate. An aperture-masking system is able to image fainter structure near to a bright object because the 'transfer function' between the object and the observables, especially the closure phases, is more stable and so easier to calibrate than with AO.

A disadvantage of NRM over AO is that the mask blocks out most of the light (in many cases 99% or more) and only short exposure times can be used. As a result, most aperture-masking results have been on relatively bright targets. If aperture masking is used in conjunction with AO, the AO will reduce the level of piston phase difference and so the exposure time can in principle be extended to many t_0 and thereby allow aperture masking to be used on fainter targets (Bernat *et al.*, 2010).

Most of the design considerations which apply to long-baseline interferometers apply also to NRM setups, with the practical constraint that NRM instruments typically have to fit in with other instruments on the same telescope and so cannot be too large or complex. Array layout is more flexible

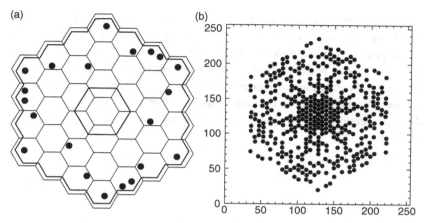

Figure 4.21 The mask layout for a Keck aperture masking experiment (a) and the resulting (u, v) coverage (b). From Tuthill (2012).

in NRM: in particular the number of 'telescopes' can be as large as is consistent with having a non-redundant layout. As a result it is possible for a single mask configuration to provide sufficient (u, v) coverage to make a high-quality image (see Figure 4.21), but the (u, v) coverage can also be increased by using different masks in succession or rotating a single mask.

Separate delay lines are not necessary in an NRM system because the optical delays are equalised at a gross level by pointing the telescope as a whole towards the object, and an AO system can function effectively as a fringe tracker to remove the fine errors in OPD.

The beam combination in an NRM system is by default an image-plane beam combination in a Fizeau mode (see Section 4.7.1). A disadvantage of this form of beam combination is that it couples the array layout to the beam-combination layout, and if spectral dispersion is required then only linear arrays can be used. A number of initiatives to do 'pupil remapping' are under way to overcome this problem (Perrin *et al.*, 2006a; Jovanovic *et al.*, 2012).

The Large Binocular Telescope (LBT) can be considered to be a special form of an aperture-masking system, with the two telescopes acting as two large 'holes' in a virtual mask. Because the telescopes are large compared with r_0 AO is required for each telescope, together with a fringe tracker to cophase the two telescopes.

The wide field of view offered by the Fizeau mode means that cophasing the two telescopes on a bright star within the field should stabilise the fringes on any star within the isoplanatic patch of the atmosphere. Rotation of the baseline between the two telescopes comes from observing the target at different times of night, so that the paralactic angle rotation of telescope on its mount gives a form of 'Earth rotation synthesis'. Because there are only two apertures, closure phase techniques are not currently envisioned for use in the conventional two-telescope interferometry, but inserting an aperture mask into the beam has been proposed (Stürmer and Quirrenbach, 2012) and this would allow conventional 'dilute-aperture' interferometric imaging techniques to be employed.

5

Measurement noise

Fringe visibility measurements and derived observables such as the power spectrum and bispectrum are subject to both systematic and random errors. Chapter 3 showed how a systematic reduction in the fringe contrast can arise from the effects of atmospheric seeing. This chapter concentrates on the main sources of random errors or *noise*.

Noise on fringe parameters can arise both from the effects of atmospheric seeing and also from noise sources associated with measuring the light intensity levels in the fringe pattern. The main aim of this chapter is to derive robust 'rule-of-thumb' estimates for the noise levels in a given observation. These rules of thumb can then be used to determine which noise sources are dominant and whether random errors or systematic errors provide the fundamental limitation to accuracy in any given case.

5.1 Atmospheric noise

5.1.1 Power spectrum

As discussed in Chapter 3, the spatial and temporal wavefront phase fluctuations which occur due to atmospheric seeing cause the mean fringe contrast to decrease as the apertures go from being point-like to being comparable in size to the Fried parameter of the seeing r_0, and as the exposure times go from being infinitesimal to being comparable to the coherence time of seeing t_0. Under these conditions the fringe contrast will also fluctuate on an exposure-to-exposure basis. Exposure times and aperture sizes need to be as large as possible in order to get more light, so it is helpful to be able to quantify what the trade-off between these experimental variables and the atmospheric noise level is.

An idea for how the noise varies as a function of integration time can be obtained using the 'random-walk' model for the visibility reduction used in Section 3.3.2. In this model, the coherent flux is given by a random walk consisting of n steps so that

$$F_{ij} = F_0 \sum_{k=1}^{n} e^{i\Phi_k}, \tag{5.1}$$

where F_0 is the coherent flux in a single 'step' and Φ_k is the fringe phase at step k. The mean power spectrum is therefore given by

$$\left\langle P_{ij} \right\rangle = \left\langle |F_{ij}|^2 \right\rangle = |F_0|^2 \left\langle \sum_{k=1}^{n} e^{i\Phi_k} \sum_{l=1}^{n} e^{-i\Phi_l} \right\rangle. \tag{5.2}$$

The order of summation and averaging can be exchanged to give

$$\left\langle P_{ij} \right\rangle = |F_0|^2 \sum_{k=1}^{n} \sum_{l=1}^{n} \left\langle e^{i[\Phi_k - \Phi_l]} \right\rangle. \tag{5.3}$$

In this summation, there are n terms for which $k = l$ and $n^2 - n$ terms for which $k \neq l$. The former terms are unity, while the latter terms are random phasors, which average to zero. Thus,

$$\left\langle P_{ij} \right\rangle = n|F_0|^2, \tag{5.4}$$

which is the familiar result that a random walk of n steps has a mean-squared length which is n times the step length.

The variance of the power spectrum can be calculated from

$$\mathrm{var}(P_{ij}) = \left\langle P_{ij}^2 \right\rangle - \left\langle P_{ij} \right\rangle^2 \tag{5.5}$$

and expanding the summations as before. The second-order moment of the power spectrum is given by

$$\left\langle P_{ij}^2 \right\rangle = |F_0|^4 \sum_{k=1}^{n} \sum_{l=1}^{n} \sum_{m=1}^{n} \sum_{p=1}^{n} \left\langle e^{i[\Phi_k - \Phi_l + \Phi_m - \Phi_n]} \right\rangle. \tag{5.6}$$

The non-zero terms in this expansion occur when $k = l = m = p$ (n terms), when $k = l \neq m = p$ ($n^2 - n$ terms) and when $k = n \neq m = l$ ($n^2 - n$ terms). Thus,

$$\left\langle P_{ij}^2 \right\rangle = |F_0|^4 (2n^2 - n), \tag{5.7}$$

and so

$$\mathrm{var}(P_{ij}) = (n^2 - n)|F_0|^4. \tag{5.8}$$

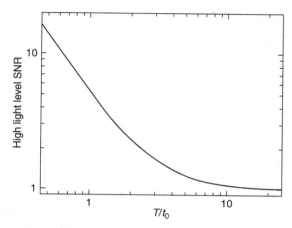

Figure 5.1 The SNR of atmospheric-noise-limited measurements of fringe amplitude as a function of exposure time. The SNR is defined as the mean-squared
visibility modulus divided by the standard deviation of the squared modulus. From
Buscher (1988a).

The 'signal-to-noise ratio' (SNR) is a common way of denoting the noise level
in a dimensionless form and the SNR of the power spectrum will be given by

$$\text{SNR}(P_{ij}) = \frac{\langle P_{ij} \rangle}{\sqrt{\text{var}(P_{ij})}} = \sqrt{\frac{1}{1 - 1/n}}. \tag{5.9}$$

Thus, the atmospheric-noise-limited SNR is infinite for small exposure times
(in practice it is limited by detection noise) and decreases towards unity for
large n. It is notable that the SNR does not decrease to arbitrarily small values but saturates at a finite value – this is not true for detection-noise-limited
observations or for atmospheric-noise-limited bispectrum measurements.

Figure 5.1 shows a plot of the SNR of atmospheric-noise-limited measurements computed using simulated temporal phase fluctuations corresponding to
a Kolmogorov model. The results agree qualitatively with the results from the
random-walk model in that the atmospheric noise SNR starts out at infinity for
small integration times and then decreases to unity for integration times much
greater than t_0.

The random-walk model can also be applied to evaluating the noise on apertures larger than r_0. This is borne out by simulations as shown in Figure 5.2,
which show that the SNR falls to unity for apertures of a few r_0 in diameter
(somewhat unexpectedly the SNR falls to slightly below unity and then rises
back towards unity, an indication that the random-walk model does not reveal
all the complexities of the more realistic models).

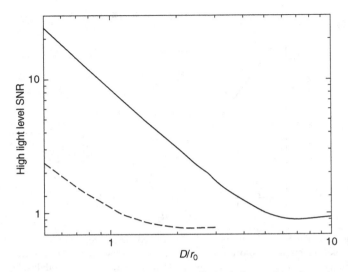

Figure 5.2 The SNR for atmospheric-noise-limited observations through apertures of different diameters (from Buscher, 1988a). The solid line corresponds to a tip–tilt-corrected aperture and the dashed line to an uncorrected aperture.

5.1.2 Bispectrum

For large aperture sizes or long integration times, the closure phase will also be subject to noise. The random-walk model again gives insight to the form the noise takes. The bispectrum measured in an integration time of nt_0 or an aperture of area nr_0^2 will be given by

$$
\begin{aligned}
T_{123} &= \frac{F_{12}}{n} \sum_{j=1}^{N} e^{i[\Phi_{1j}-\Phi_{2j}]} \frac{F_{23}}{n} \sum_{k=1}^{N} e^{i[\Phi_{2k}-\Phi_{3k}]} \frac{F_{31}}{n} \sum_{l=1}^{N} e^{i[\Phi_{3l}-\Phi_{1l}]} \\
&= \frac{F_{12}F_{23}F_{31}}{n^3} \sum_{j=1}^{N} \sum_{k=1}^{N} \sum_{l=1}^{N} e^{i[\Phi_{1j}-\Phi_{2j}+\Phi_{2k}-\Phi_{3k}+\Phi_{3l}-\Phi_{1l}]}.
\end{aligned}
\tag{5.10}
$$

The summation can be split into three sets of terms (Readhead *et al.*, 1988). There are n terms where $j = k = l$, which yield the unperturbed bispectrum. There are $3n(n-1)$ terms ('terms of the second kind') of the form $j = k \neq l$ or $j = l \neq k$ or $j \neq k = l$. These terms can be paired as complex conjugates, for example the terms for $j = q, k = q, l = s$ have the form

$$
e^{i[\Phi_{1q}-\Phi_{1s}-\Phi_{3q}+\Phi_{3s}]}
\tag{5.11}
$$

while the terms for $j = s, k = s, l = q$ have the form

$$
e^{-i[\Phi_{1q}-\Phi_{1s}-\Phi_{3q}+\Phi_{3s}]}.
\tag{5.12}
$$

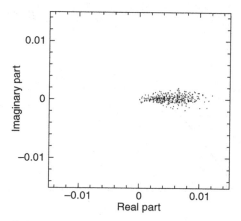

Figure 5.3 Sample bispectrum values plotted on an Argand diagram. The values were measured using aperture masking on a bright point source (α Boo) with sub-r_0 subapertures. The scatter is larger in the direction parallel to the mean bispectrum value than in the perpendicular direction, suggesting that the exposure time used was much greater than t_0. From Haniff and Buscher (1992).

The sum of these two terms has no imaginary component and so adds no noise in the direction perpendicular to the mean bispectrum. As a result, the main effect of these terms is to cause fluctuations in the amplitude of the bispectrum and not of the phase (though they can cause the phase to be in error by 180°). There are $n(n - 1)(n - 2)$ terms of the third kind where $j \neq k \neq l$; these have random phase and so add noise equally to the directions parallel and perpendicular to the mean bispectrum vector.

The above model predicts that for small n there will be more noise in the direction parallel to the mean bispectrum (primarily caused by the terms of the second kind) than in the direction perpendicular to the mean bispectrum (which would affect the phase). This is indeed what is seen in real data as shown in Figure 5.3.

If the bispectrum is averaged over N_{exp} exposures so that the noise on the averaged bispectrum is small, then the error on the closure phase, i. e. the phase of the bispectrum, will be approximately

$$\sigma = \frac{\sigma_\perp}{|\langle T_{123} \rangle| \sqrt{N_{exp}}}, \tag{5.13}$$

where $\sigma_\perp$ is the standard deviation of the noise in the direction perpendicular to the direction of the mean bispectrum. It is convenient to define the phase error σ_ϕ for a single exposure

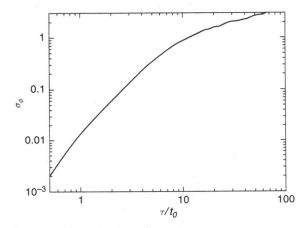

Figure 5.4 The phase error σ_ϕ on the bispectrum as a function of exposure time τ for atmospheric-noise-limited measurements. From Buscher (1988a).

$$\sigma_\phi = \frac{\sigma_\perp}{|\langle T_{123}\rangle|}. \tag{5.14}$$

Note that it is possible for σ_ϕ to be much larger than 2π but still have meaning: if the bispectrum is averaged over $N_{\exp}$ exposures, then the noise on the averaged bispectrum will be a factor $\sqrt{N_{\exp}}$ smaller than the noise on a single frame, and if sufficient exposures are averaged such that $\sigma_\perp / \sqrt{N_{\exp}} \ll |\langle T_{123}\rangle|$ then the noise on the phase of the averaged bispectrum will be $\sigma_\phi / \sqrt{N_{\exp}}$. Thus, the phase error σ_ϕ can be interpreted in terms of the number of exposures needing to be averaged in order to get a given phase standard deviation.

In the case of the random-walk visibility model, $\sigma_\perp \propto \sqrt{N^3}$ for large N while $\langle T_{123}\rangle \propto N$, so the phase error will increase as $\sqrt{N}$. Figures 5.4 and 5.5 show that this model is a good guide to the results of more accurate simulations, with the error rising as $\sqrt{\tau}$ for long exposure times and as D for large aperture sizes. For shorter exposure times and smaller apertures the phase error decreases more steeply, and the phase error remains below a radian for integration times of nearly $10t_0$. This can be explained in terms of the random-walk model as being due to the fact that the only noise terms which give rise to a finite phase error are of the third kind, and there are no such terms for $n < 3$.

5.2 Detection noise

In addition to atmospheric noise, the fringe measurements are affected by *detection noise*, which results when the light is converted into an electronic signal. Figure 5.6 shows the typical detection process schematically.

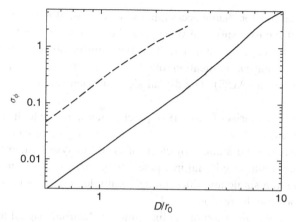

Figure 5.5 The phase error σ_ϕ on the bispectrum as a function of aperture diameter for atmospheric-noise-limited measurements. The dashed line is for no adaptive optics correction and the solid line is for tip–tilt correction. From Buscher (1988a).

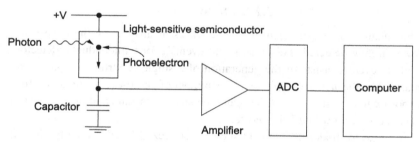

Figure 5.6 The signal-processing chain for a single pixel of a photoelectric detector. Light enters as photons on the left and exits as a digitised electrical signal on the right.

Detectors typically make use of the photoelectric effect in which incoming photons create free charge carriers in a material when it is exposed to light. A promising alternative detection mechanism makes use of superconducting materials in which the incoming photons break Cooper pairs rather than generating photoelectrons, but the noise issues present in the readout process are similar to those modelled here for photoelectric devices. The photoelectic detection material is usually a semiconductor, with semiconductors with different bandgaps being used at different wavelengths, for example silicon at visible wavelengths and mercury–cadmium–telluride (HgCdTe) at near-infrared wavelengths.

The photoelectric current is typically accumulated over the course of an exposure: this is represented in the diagram as the charging of a capacitor, but the mechanism for charge accumulation varies with different technologies. At the end of the exposure, the voltage on the capacitor is read to determine the amount of charge accumulated. This typically involves several stages of charge and/or voltage amplification followed by digitisation using an analogue-to-digital converter (ADC). The digital signal is then sent to a computer for processing.

Most detectors consist of a matrix of pixels, each sensitive to the light falling on a small area, so that a two-dimensional intensity pattern can be read out. In order to reduce the amount of electronics in the system, these pixels are typically time-multiplexed: multiple pixel charge accumulators are connected one at a time to a single amplifier, though multiple amplifiers may operate in parallel to increase the readout speed.

The process of conversion of the light into an electronic signal introduces a number of sources of noise. The most important noise sources are discussed in the following subsections.

5.2.1 Photon noise

The quantum nature of electromagnetic waves means that, at the lowest light levels, light is detected as discrete 'photo-events'. For semiconductor detectors, a photo-event consists of the generation of a single electron/hole pair in the material by the arrival of a photon. The number of photo-events fluctuates from exposure to exposure even if the light intensity is constant, and this fluctuation is called *shot noise* or *photon noise*.

In order to model phenomena such as 'squeezed light' correctly, the full quantum field theory of light, which treats the electromagnetic field as being quantised into photons from the start, is necessary. However, all the scenarios encountered in astronomical interferometry can be modelled to high accuracy using a simpler 'semiclassical' approximation (Mandel *et al.*, 1964), in which light waves propagate following classical laws (i. e. Maxwell's equations), but are quantised on interaction with a detector.

In the semiclassical model, the number of photo-events in a given pixel in a given exposure follows a Poisson distribution with mean Λ_p given by

$$\Lambda_p = \int_{\text{exposure}} \int_{\text{bandpass}} \iint_{\text{detector surface}} \frac{\eta_p(x, y, \nu, t)I(x, y, \nu, t)}{h\nu} \, dx \, dy \, d\nu \, dt,$$

(5.15)

where $I(x, y, \nu, t)$ is the classical specific light intensity (energy per unit area per unit time per unit frequency) falling on location (x, y) at frequency ν and time t, $\eta_p(x, y, \nu, t)$ is the *quantum efficiency* (QE) of the detector pixel as a function of location, frequency and time and h is Planck's constant (so $h\nu$ is the energy of a photon at frequency ν). The QE is a value between 0 and 1 which relates the number of photo-events actually occuring in a detector to the maximum number of photo-events which could be recorded by an ideal device with the same incident light intensity. The mean of the distribution Λ_p will be referred to hereafter as the *integrated classical intensity* for a given pixel and exposure but it should be remembered that Λ_p is expressed in units of photo-events rather than joules.

The number of photo-events recorded in an exposure N_{phot} will have a mean of Λ_p and a standard deviation of

$$\sigma_p = \sqrt{\Lambda_p}. \tag{5.16}$$

For $\Lambda_p \gtrsim 10$ the Poisson distribution can be approximated by a Gaussian distribution with the same standard deviation.

5.2.2 Readout noise

The electronics used to amplify the accumulated photocurrent signal in the detector will introduce additional noise and this is characterised as a *readout noise* (also known as *read noise*). This noise usually has a Gaussian distribution, which is typically independent of the signal level and uncorrelated from pixel to pixel (noise which is correlated from pixel to pixel is often called *pattern noise* and in good detectors is kept at levels well below the readout noise). Values for the readout noise can range from 100 electrons to less than an electron in devices which have near-noiseless amplification.

5.2.3 Background noise

Background noise arises from signals arriving at the detector from sources other that the object being observed. These sources include stray light, sky emission or, at infrared wavelengths longward of about 2 μm, by thermal emission from the optics. These arise from causes independent of the detector, but are included in 'detection noise' because the noise characteristics are similar to other noise processes arising in the detector. Thermally induced 'dark currents' in the detector are indistinguishable from photoelectrons and can also be included as a form of background.

The noise arises primarily from the shot noise of the additional radiation or dark current received during an exposure, and so can be modelled by including a constant offset in the expression for Λ_p given in Equation (5.15). Background noise differs from 'normal' photon noise in that its variance is independent of the brightness of the target object, and it differs from readout noise in that its variance increases as exposure time is increased whereas the readout noise variance is usually independent of the exposure time.

5.2.4 Electron amplification noise

Some detectors use a process known as 'impact ionisation' in semiconductors to increase the number of charge carriers generated by a photo-event. In avalanche photodiodes, the photocurrent is amplified before charge accumulation and in electron-multiplying CCDs the charge is amplified in an 'amplification register' after charge accumulation. In both cases, the amplification introduces extra noise, called amplification noise or excess noise. These noise sources can be approximately modelled as increasing the level of photon noise by a fixed factor.

5.3 Alternative fringe detection methods

The detection process described above assumes that the interfering light beams are brought directly together on a detector to form fringes. In radio interferometry quite different technologies are used to form fringes and this has a large impact on the form of the detection noise. Almost all radio interferometers amplify the incoming signal at each telescope and also use heterodyne fringe detection. Although heterodyne techniques have been used at mid-infrared wavelengths, neither of these technologies is in use at shorter wavelengths. The reasons for this are discussed below.

5.3.1 Heterodyne interferometry

In a *heterodyne* system, the incoming electromagnetic wave is converted before beam combination. The conversion is achieved by interfering the incoming ware with a locally generated signal at a similar frequency, the so-called 'local oscillator'. At optical wavelengths this local oscillator would likely be a laser. The interference takes place on a fast detector, which converts the rapid oscillations in intensity from the 'beats' between the astronomical signal and the local oscillator into an electrical voltage. This electrical signal is then amplified and transmitted to a central location where fringes are formed electronically in a mathematically equivalent way to a conventional optical beam combiner. The interference signal can be derived by summing the instantaneous voltages from each pair of telescopes and squaring the result or alternatively by multiplying the voltages.

The interference signal derived using a heterodyne interferometer is identical (apart from the noise) to the signal, which would be seen in an equivalent direct detection (*homodyne*) interferometer, but with the advantage that the signal can be amplified before transmission and the signals can be transmitted via cables rather than light pipes. Heterodyne technology is used in the Berkeley ISI interferometer (Hale *et al.*, 2000), which operates at a wavelength of 11 μm, but has not been used at shorter wavelengths due to a number of technical difficulties as explained below.

5.3.2 Optical amplification

An alternative to heterodyne fringe detection is to use direct detection of the fringes but to use an optical amplifier to increase the intensity of the light beams before beam combination. This would allow the light from each telescope to be split multiple ways to form pairwise interference fringes with multiple other telescopes without the accompanying loss of signal level in each interference pattern. Such an amplifier at optical wavelengths would likely be in the form of a laser, in other words a cavity containing 'pumped' atoms, which coherently amplifies the incoming signal. No such *pre-amplifier* technology has been used in an optical interferometer to date.

5.3.3 Bandwidth limitations

Both pre-amplifiers and heterodyne systems allow the signal to be amplified at the telescope before transmission to the beam combiner, and therefore overcome losses in the beam transmission and beam combination system. However, there are a number of both technological and fundamental problems, which mean that, at wavelengths shortwards of about 5 μm, these technologies are unlikely to see the kind of near-ubiquitous use seen in radio-frequency interferometers.

The first disadvantage of these techniques is that with current technology they work over relatively restricted bandwidths. Heterodyne technology in particular has a spectral bandwidth which is less than or equal to the highest frequency of intensity fluctuation that the detector can respond to. For the best modern detectors, this is of order a few gigahertz. This may sound like a significant bandwidth but a few gigahertz corresponds to a fractional bandwidth $\Delta \nu / \nu$ of order 10^{-5} at optical frequencies. Compared with homodyne systems whose fractional bandwith can easily exceed 10^{-2}, a heterodyne system can suffer a large loss in sensitivity.

A way around the bandwidth limitation is to disperse the light spectrally and run many systems in parallel at different frequencies. This is becoming more feasible with heterodyne systems because of the availability of optical-frequency combs, which can provide local oscillators at hundreds of different frequencies simultaneously (Ireland and Monnier, 2014). In this case the inherently narrow bandwidth of the heterodyne system can be turned to advantage when observing narrow spectral lines in the objects of interest.

5.3.4 Quantum noise limits

A more fundamental limitation arises from the fact that both heterodyne systems and pre-amplifier systems include a 'coherent' amplification process. All amplifiers of electromagnetic radiation are fundamentally limited by an irreducible level of quantum noise. Amplifiers which are coherent in the sense that they preserve the phase information in the electromagnetic field will have a noise level which can be much higher than the photon noise associated with non-phase-preserving amplification processes such as those which occur in photomultipliers.

Any coherent amplification process can be shown to have a noise level which is equivalent to one photon per second per hertz of amplifier bandwidth at the input to the amplifier (Caves, 1982). This assumes a single-spatial-mode amplifier as would be appropriate for an interferometer: the more general result is that there is one noise photon per amplified 'mode' of the electromagnetic field. The noise is unavoidable and can be interpreted as arising from the sensitivity of any amplifier of electromagnetic fields to zero-point quantum fluctuations or 'vacuum' fluctuations.

At optical wavelengths, this noise source will be significant. For example, an optical amplifier with a bandwidth of 1 GHz will have an equivalent noise level of 10^9 noise photons per second at its input. This can be compared to typical photon rates of perhaps 10^6 photons per second in an optical interferometer observing a relatively bright object using a bandpass of many hundreds of gigahertz.

A more definitive idea of the effect of this noise at any wavelength can be obtained by comparing the amplification noise to the highest possible signal levels expected from typical astronomical sources (Prasad, 1994). These sources can be usually be approximated as thermal blackbody radiators at some temperature T and so the number of photons per mode of the radiation at frequency ν from these sources will be given by

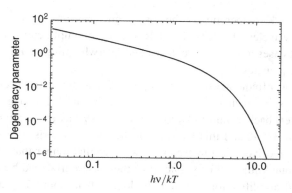

Figure 5.7 The photon degeneracy parameter (photons per unit frequency per unit time in a single spatial mode) for thermal radiation at frequency ν from a black body at temperature T.

$$\delta = \frac{1}{e^{h\nu/(kT)} - 1}, \tag{5.17}$$

where h is Planck's constant and k is the Boltzmann factor (Mandel, 1961).

This 'photon degeneracy parameter' represents the maximum number of photons per unit bandwidth per unit time which can be injected into a single-mode optical fibre by a telescope observing a thermal radiator at temperature T. Note that this maximum is reached when the angular size of the radiator is larger than the diffraction limit of the telescope, which is unlikely to occur for most astronomical objects of interest to interferometry.

Since any amplifier will be subject to a noise input of one photon per unit bandwidth per unit time, then δ is the maximum SNR of the output of an amplification system when observing a source at temperature T. It can be seen from Figure 5.7 that δ falls below unity at frequencies ν such that $h\nu/kT > \log 2 \approx 0.69$, and drops rapidly for $h\nu/kT \gg 1$. This means that at high frequencies the signal will be swamped by amplifier noise even when using the largest possible telescope and observing the largest possible object. For example, the radiation from an object at a temperature of 6000 K (typical for a star like the Sun) has a degeneracy parameter at a wavelength of 500 nm of $\delta \approx 8.3 \times 10^{-3}$ and, under these circumstances, optical amplification systems are of less than no use.

The reason why using pre-amplification and heterodyne systems make sense in radio interferometry is that the much lower frequencies involved mean that the photon degeneracy parameter is significantly higher. For example, at a wavelength of 1 cm a 6000 K object has a photon degeneracy parameter of more than 4000. Even the coldest objects in the Universe, which are typically at around the 3 K temperature of the cosmic microwave background, will have $\delta > 1$ at this wavelength. Thus, the SNR penalty for amplification is negligible in almost all cases at radio frequencies and overwhelming in almost all cases at optical frequencies.

The crossover frequency at which amplification technologies become a practical option is somewhere in the mid- to far-infrared. For a ground-based interferometer, background radiation from the sky has an effective temperature of about 300 K and this background flux will comprise the majority of the radiation received by a telescope at wavelengths longer than about 3 μm. This background flux will be greater than the noise introduced by a quantum-noise-limited amplifier for wavelengths longer than about 70 μm and so at these wavelengths the noise penalty for using an amplifier may be outweighed by other practical advantages such as the ability to easily split the signal many ways.

These practical advantages might extend the usefulness of heterodyne interferometry to wavelengths as short as 10 μm in interferometers with many telescopes (Ireland and Monnier, 2014), but it is unlikely that heterodyne systems will be useful at much shorter wavelengths because of the steep decrease in the photon degeneracy parameter with increasing frequency.

It should be noted that frequency upconversion, which is a process where the incoming photons are converted into photons at a higher frequency (Ceus *et al.*, 2013), is not subject to the amplification noise described above. Upconversion is useful because it can allow the use of optical fibres and detectors at frequencies where these technologies may be more mature than the same technologies at lower frequencies, but it does not result in a gain in the number of photons and so is not subject to the amplification noise of a true amplifier.

5.3.5 Intensity interferometry

An alternative but related form of fringe detection is intensity interferometry (Hanbury-Brown and Twiss, 1956). In this technique, the fluctuations in intensity of the light from a star are recorded using a fast detector at each telescope. The intensity signals are transmitted to a central signal combiner that measures the correlations in these fluctuations between telescopes, which are related to the modulus squared of the coherent flux on the projected baseline between the telescopes. The variation of the modulus of the coherent flux with baseline can be used to measure the diameters of stars (Hanbury-Brown *et al.*, 1967) and the separation of close binaries (Hanbury-Brown *et al.*, 1970).

An advantage of using an intensity interferometer over more conventional 'amplitude' interferometers is its insensitivity to mechanical or atmospheric delays. This means that large but imprecise telescopes such as those present in the telescope arrays used for detecting Cherenkov radiation can be used as light collectors for intensity interferometry (Bohec and Holder, 2006).

However, even with large collectors intensity interferometry has so far been limited to the observation of bright stars. The reason for this is that, like heterodyne interferometry, the SNR of intensity interferometry is limited by the photon degeneracy parameter of the radiation, and for most astronomical objects this much less than unity at optical wavelengths as shown in Section 5.3.4.

An additional reason why intensity interferometry has not been developed as much as amplitude interferometry is that, although it may be possible to measure closure-phase-type quantities using intensity interferometry on three or more telescopes (Malvimat *et al.*, 2014), in most practical cases intensity interferometry is restricted to measurement of visibility-modulus information, which, as demonstrated in Section 9.6.2, severely reduces its applicability to the imaging of complex sources.

5.4 The interferogram

The following analysis assumes that the fringes are formed in by direct detection of un-amplified optical beams. The effects of detection noise on the measurement of fringe parameters can be understood from a simple model of the fringe pattern and how it is analysed. Different types of fringe pattern and different fringe analysis methods will be presented in Section 8.5, but the results from the simple model used here can be straightforwardly extended to apply to more complex practical examples.

The model of the detected fringe pattern consists of a one-dimensional sinusoidal light-intensity pattern sampled by a set of N_{pix} evenly spaced pixels. The integrated classical intensity on pixel p in a given exposure or 'interferogram' is given by

$$\Lambda_p = \frac{1}{N_{\text{pix}}} \left(\overline{N}_{\text{phot}} + \text{Re} \left\{ F_{ij} e^{2\pi i s_{ij} p} \right\} \right), \tag{5.18}$$

where $\overline{N}_{\text{phot}}$ is the classical light intensity integrated over all pixels and expressed as a number of photons, F_{ij} is the coherent flux of the fringe expressed in photon units and s_{ij} is the spatial frequency of the fringes. It is assumed that there is an integral number of fringes across the interferogram, i. e. that $s_{ij} N_{\text{pix}}$ is an integer.

The detected values for the integrated intensity in pixel p can be written

$$i_p = \Lambda_p + n_p, \tag{5.19}$$

where n_p is the noise value on pixel p for the given exposure. Note that Equation (5.19) implies that i_p and n_p are scaled to be in the same units as Λ_p, i.e. photons.

For simplicity, n_p will be assumed to have zero mean. The noise is assumed to be uncorrelated between pixels: this is an accurate approximation under most situations of practical interest. An additional assumption is that the noise has a Gaussian probability distribution, which is accurate in situations where read noise or background noise are the dominant sources of noise, or in situations where photon noise is dominant and Λ_p is larger than about 10.

The noise on pixel P is assumed to have a standard deviation σ_p that does not vary significantly from pixel to pixel, which is the case for many read-noise-limited detectors and for photon-noise-limited measurements if the fringe contrast is low so that $|F_{ij}| \ll \overline{N}_{\text{phot}}$. Results from this signal-independent Gaussian noise model give a good idea of the effects of noise in most situations of practical interest.

5.5 Noise on fringe parameters

5.5.1 Estimating the coherent flux

To derive a reliable estimate for a parameter such as the coherent flux F_{ij} from the noisy data i_p is a problem in statistical inference. Section 8.5 discusses the use of statistical 'estimators' in such problems and introduces a number of estimators for the coherent flux. For the purposes of this analysis the estimator that will be used is the discrete Fourier transform (DFT) of the intensity pattern at frequency s_{ij}.

The DFT estimator $\hat{F}_{ij}$ for the quantity F_{ij} is given by

$$\hat{F}_{ij} = 2 \sum_{p=0}^{N_{\text{pix}}-1} i_p e^{-2\pi i s_{ij} p}. \tag{5.20}$$

Substituting Equations (5.19) and (5.18) into Equation (5.20) gives

$$\hat{F}_{ij} = F_{ij} + n_{ij}, \tag{5.21}$$

where n_{ij} is given by

$$n_{ij} = 2 \sum_{p=0}^{N_{\text{pix}}-1} n_p e^{-2\pi i s_{ij} p}. \tag{5.22}$$

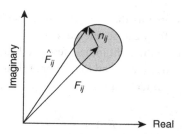

Figure 5.8 Argand diagram representation of the estimator $\hat{F}_{ij}$ for the fringe coherent flux as a sum of the true coherent flux F_{ij} and the complex noise n_{ij}. It can be seen that the complex noise gives rise to both amplitude and phase errors in the estimate.

Thus, in the presence of noise $\hat{F}_{ij}$ is offset from F_{ij} by a complex noise vector n_{ij} as shown in Figure 5.8. This complex noise causes both the amplitude and the phase of the measured coherent flux to deviate from the true value.

The sum in equation (5.22) is a two-dimensional random walk in the complex plane with a root-mean-square (RMS) 'step size' of σ_p. As the number of pixels becomes large, the probability distribution of n_{ij} will tend towards a two-dimensional circularly symmetric Gaussian. The variance of n_{ij} will be given by

$$\text{var}(n_{ij}) \equiv \sigma_{ij}^2 = 4 \sum_p \sigma_p^2. \tag{5.23}$$

For photon-noise-limited measurements,

$$\sigma_p = \sqrt{\Lambda_p}, \tag{5.24}$$

and so

$$\sigma_{ij} = 2 \sqrt{\overline{N}_{\text{phot}}}, \tag{5.25}$$

where

$$\overline{N}_{\text{phot}} = \sum_p \Lambda_p \tag{5.26}$$

is the integrated classical intensity summed over all pixels. The noise on the complex flux measurement is therefore independent of the number of pixels over which the light is spread.

For read-noise-limited measurements, assuming that the read noise is the same on all pixels so that $\sigma_p = \sigma_{\text{read}}$, then

$$\sigma_{ij} = 2\sigma_{\text{read}} \sqrt{N_{\text{pix}}}, \tag{5.27}$$

and so the noise gets worse as the light is spread out over more pixels. As a consequence, designs for beam combiners which feed read-noise-limited detectors try to concentrate the light into as few pixels as possible while those for photon-noise-limited detectors tend to be more relaxed in their pixel usage.

The SNR of the measurement is given by

$$\text{SNR}(\hat{F}_{ij}) = \frac{|\langle \hat{F}_{ij} \rangle|}{\sigma_{ij}}. \tag{5.28}$$

In the photon-noise-dominated regime the SNR is given by

$$\text{SNR}(\hat{F}_{ij}) \approx \frac{1}{2} |V_{ij}| \sqrt{N_{\text{phot}}}, \tag{5.29}$$

where $V_{ij} = F_{ij}/\overline{N}_{\text{phot}}$ is the fringe visibility, while in the read-noise-dominated regime the SNR is given by

$$\text{SNR}(\hat{F}_{ij}) \approx \frac{V_{ij}\overline{N}_{\text{phot}}}{2\sigma_{\text{read}}\sqrt{N_{\text{pix}}}}. \tag{5.30}$$

Comparing Equations (5.29) and (5.30), it can be seen that, for photon-noise-limited measurements, the decrease in SNR caused by a given percentage loss in fringe contrast is larger than would be caused by the same percentage loss in photon rate, while the effects of visibility loss and photon loss are equal in the case of read-noise-limited measurements. These differences in the sensitivity to visibility loss versus photon loss have implications when optimising the SNR of a fringe measurement, and these implications are discussed further in Chapter 6.

5.5.2 Power spectrum bias and noise

If the object being observed is faint, the SNR in a single exposure is often less than unity. An obvious way to increase the SNR is to average multiple exposures, since with exposure times measured in milliseconds, thousands of exposures can be taken in a few seconds. However, the estimator $\hat{F}_{ij}$ is unsuitable for such averaging because its phase will change randomly from exposure to exposure, and so the average will tend towards zero amplitude.

As the power spectrum and bispectrum are more stable in the presence of atmospheric phase disturbances, averaging these quantities over many exposures is possible. The effects of measurement noise on the power spectrum can be understood in terms of the simple model with Gaussian noise used above.

An initial guess at an estimator for the power spectrum would be

$$\hat{P}_{ij,\text{biased}} = |\hat{F}_{ij}|^2. \tag{5.31}$$

However, since this estimator is to be averaged over multiple exposures it is important that the average over a large number of exposures converges on the true power spectrum value. Substituting Equation (5.21) into Equation (5.31) and taking the average gives

$$\left\langle \hat{P}_{ij,\text{biased}} \right\rangle = \left\langle |F_{ij}|^2 \right\rangle + 2\text{Re}\left\{ \left\langle F_{ij}n^* \right\rangle \right\} + \left\langle |n|^2 \right\rangle$$

$$= \left\langle |F_{ij}|^2 \right\rangle + \sigma_{ij}^2, \tag{5.32}$$

since the cross term $F_{ij}n^*$ has random phase and therefore averages to zero. Thus, averaging the value of the estimator $\hat{P}_{\text{biased}}$ over a large number of frames does not converge on the noise-free value and so this estimator is, as its name suggests, biased.

If the value of σ_{ij} is known, then an unbiased estimator can be constructed:

$$\hat{P}_{ij} = \left| \hat{F}_{ij} \right|^2 - \sigma_{ij}^2 \tag{5.33}$$

such that

$$\left\langle \hat{P} \right\rangle = \left\langle |F_{ij}|^2 \right\rangle. \tag{5.34}$$

The variance of the power spectrum estimator $\hat{P}_{ij}$ can be estimated using the standard formula

$$\sigma^2(x) = \left\langle x^2 \right\rangle - \langle x \rangle^2, \tag{5.35}$$

where $\sigma^2(x)$ is the variance of a quantity x. Substituting Equations (5.31), (5.33) and (5.21) into Equation (5.35), making use of the result that a circularly symmetric complex Gaussian distribution has a fourth-order moment $\left\langle |n|^4 \right\rangle = 2\sigma_{ij}^4$, and eliminating terms with zero mean gives

$$\sigma(\hat{P}_{ij})^2 = \text{var}(|F_{ij}|^2) + 2\left\langle |F_{ij}|^2 \right\rangle \sigma_{ij}^2 + \sigma_{ij}^4, \tag{5.36}$$

where $\text{var}(|F_{ij}|^2)$ is the variance of the squared modulus of the classical fringe amplitude due to atmospheric noise.

The SNR of the power spectrum is therefore given by

$$\text{SNR}(\hat{P}_{ij}) = \frac{\left\langle |F_{ij}|^2 \right\rangle}{\sqrt{\text{var}(|F_{ij}|^2) + 2\left\langle |F_{ij}|^2 \right\rangle \sigma_{ij}^2 + \sigma_{ij}^4}}. \tag{5.37}$$

This expression gives the SNR of a single exposure: the SNR of the average of N_{exp} independent exposures is $\sqrt{N_{\text{exp}}}$ higher than the single-exposure SNR. More general expressions for the noise on the power spectrum, which deal accurately with the combined effects of read noise and photon noise are given in Gordon and Buscher (2012), but the above expression gives a good idea of the SNR in most scenarios of practical interest.

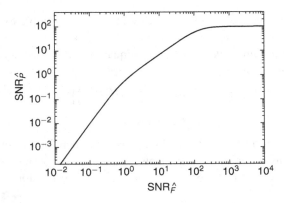

Figure 5.9 The SNR of the power spectrum as a function of the SNR of the coherent flux. The atmospheric-noise-dominated SNR has been assumed to be of order 100.

The expression for the SNR of the power spectrum given in Equation (5.37) can be split into three different regimes as evident from Figure 5.9. At the very highest light levels where the variation of the power spectrum due to atmospheric wavefront fluctuations dominates over the detection noise, i.e. $\mathrm{var}(|F_{ij}|^2)/\langle|F_{ij}|^2\rangle^2 \gg \sigma_{ij}^2/\langle|F_{ij}|^2\rangle$, the SNR is independent of the light level:

$$\mathrm{SNR}(\hat{P}_{ij}) \approx \frac{\langle|F_{ij}|^2\rangle}{\sqrt{\mathrm{var}(|F_{ij}|^2)}}. \tag{5.38}$$

This is the atmospheric-noise-dominated regime, and the SNR as a function of integration time and aperture diameter is given in Figures 5.1 and 5.2, respectively. At lower light levels where the detection noise is larger than the atmospheric noise but where $\langle|F_{ij}|^2\rangle \gg \sigma_{ij}^2$, the SNR of the power spectrum is linearly proportional to the SNR of the coherent flux estimator given in Equation (5.28)

$$\mathrm{SNR}(\hat{P}_{ij}) \approx \sqrt{\frac{\langle|F_{ij}|^2\rangle}{2\sigma_{ij}^2}} \approx \frac{\mathrm{SNR}(\hat{F}_{ij})}{\sqrt{2}}, \tag{5.39}$$

while at lower light levels where $\langle|F_{ij}|^2\rangle \ll \sigma_{ij}^2$ it varies quadratically with $\mathrm{SNR}(\hat{F}_{ij})$

$$\mathrm{SNR}(\hat{P}_{ij}) \approx \frac{\langle|F_{ij}|^2\rangle}{\sigma_{ij}^2} \approx \left(\mathrm{SNR}(\hat{F}_{ij})\right)^2. \tag{5.40}$$

The quadratic fall-off in SNR at the lowest light levels means that there is a strong cut-off in the faintness of objects, which can be observed in practice. For example, in order to obtain an averaged power spectrum value with an SNR of 10 when $\text{SNR}(F_{ij}) = 0.3$, approximately 10^4 exposures need to be averaged. To obtain the same SNR at a lower light level where $\text{SNR}(F_{ij}) = 0.1$, approximately 10^6 exposures need to be averaged. If an exposure is taken every 10 ms then the former observation will take less than 2 min, while the latter will take nearly 3 h. In 3 h the (u, v) coordinate sampled by the pair of telescopes will have changed significantly due to Earth rotation, and so averaging the power spectrum over 3 h will be fraught with difficulty.

5.5.3 Bispectrum noise

In order to understand the noise on the bispectrum, the model of the fringe pattern from Section 5.4 needs to be extended to allow the measurement of the visibility at three spatial frequencies. The model used in this section will be of an interferometer consisting of three collectors and a set of three pairwise beam combiners. These combiners form fringe patterns on different detectors allowing independent measurements of the object visibility at spatial frequencies u_{12}, u_{23} and u_{31}. Some beam combiners instead multiplex all three fringe patterns in a single interference pattern as explained in Section 4.7.2, but the results derived using a pairwise model are a good guide to the results in more complex scenarios.

The coherent fluxes F_{12}, F_{23} and F_{31} can be estimated using Equation (5.20) and the three estimators are denoted $\hat{F}_{12}$, $\hat{F}_{23}$ and $\hat{F}_{31}$ respectively. A bispectrum estimator can be constructed from these as simply

$$\hat{T}_{123} = \hat{F}_{12}\hat{F}_{23}\hat{F}_{31}. \tag{5.41}$$

Substituting Equation (5.21) into Equation (5.41) gives

$$\begin{aligned} \hat{T}_{123} &= (F_{12} + n_{12})(F_{23} + n_{23})(F_{31} + n_{31}) \\ &= F_{12}F_{23}F_{31} \\ &\quad + F_{12}F_{23}n_{31} + F_{23}F_{31}n_{12} + F_{31}F_{12}n_{23} \\ &\quad + F_{12}n_{23}n_{31} + F_{23}n_{31}n_{12} + F_{31}n_{12}n_{23} \\ &\quad + n_{12}n_{23}n_{31}. \end{aligned} \tag{5.42}$$

The mean value of $\hat{T}_{123}$ is the noise-free object triple product $\langle F_{12}F_{23}F_{31} \rangle$, since the rest of the terms in the expression have zero mean. Thus, the estimator $\hat{T}_{123}$ is an unbiased estimate of T_{123}.

The variance of $\hat{T}_{123}$ can be calculated by combining Equations (5.35) and (5.42). Eliminating the zero-mean terms and assuming that the standard

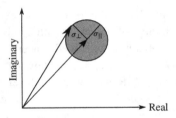

Figure 5.10 Schematic diagram of the noise on the bispectrum.

deviation of the noise is the same for all the fringe patterns, $\sigma_{12} = \sigma_{23} = \sigma_{31} = \sigma$ gives

$$\sigma(\hat{T}_{123})^2 = \sigma(F_{12}F_{23}F_{31})^2$$
$$+ \left(\langle |F_{12}F_{23}|^2 \rangle + \langle |F_{23}F_{31}|^2 \rangle + \langle |F_{31}F_{12}|^2 \rangle \right) \sigma^2$$
$$+ \left(\langle |F_{12}|^2 \rangle + \langle |F_{23}|^2 \rangle + \langle |F_{31}|^2 \rangle \right) \sigma^4 + \sigma^6. \qquad (5.43)$$

If the noise on each of the fringe coherent flux measurements has a circularly symmetric distribution in the complex plane, then it is straightforward to show that the noise on the bispectrum will also have a circularly symmetric distribution as depicted in Figure 5.10. Since the variance of a circularly symmetric noise distribution is given by $\sigma^2 = \sigma_{\perp}^2 + \sigma_{\parallel}^2 = 2\sigma_{\perp}^2$, then $\sigma_{\perp}^2$ is half the value given in Equation (5.43).

Assuming that the seeing-induced fluctuations in the fringe amplitudes are small so that $\mathrm{var}(F_{12}F_{23}F_{31})$ is small, $\left| \langle F_{ij} \rangle \right|^2 \approx \langle |F_{ij}|^2 \rangle$ and so on, then Equation (5.43) can be used to show that there will be two different SNR regimes, similar to those for the power spectrum.

When the SNR on all the baselines is sufficiently high that $\langle |F_{ij}|^2 \rangle \gg \sigma_{12}^2$ and so on, then

$$\sigma_\phi^2(T_{123}) \approx \sigma_\phi^2(F_{12}) + \sigma_\phi^2(F_{23}) + \sigma_\phi^2(F_{31}), \qquad (5.44)$$

where $\sigma_\phi^2(F_{12})$ is a phase error on the coherent flux F_{ij} defined analagously to the phase error given in Equation (5.13).

Thus, in high-SNR conditions the phase errors of the component coherent fluxes add in quadrature to give the phase error on the closure phase (as might be expected if the component phases were simply added to give the closure phase).

When the SNR is low on all baselines,

$$\sigma_\phi^2(T_{123}) \approx \sigma_\phi^2(F_{12})\sigma_\phi^2(F_{23})\sigma_\phi^2(F_{31}); \qquad (5.45)$$

therefore, in low-SNR conditions, the bispectrum phase error scales roughly as the cube of the phase error of a single coherent flux. In other words, there is an even more severe cut-off for using the bispectrum at low light levels than there is to using the power spectrum at these light levels.

5.6 Comparison of noise levels

The power spectrum and bispectrum are subject to a combination of atmospheric noise, detection noise and systematic calibration errors. The question as to which of these is most likely to be important is an obvious one when planning interferometric observations.

For sufficiently bright objects the atmospheric noise will dominate over detection noise. However, after incoherent averaging over relatively short periods the atmospheric noise can become less important than systematic errors such as calibration errors. To take an example, using a relatively long exposure time of $3t_0$ will give an SNR for the power spectrum of approximately 1.5 for a single exposure (see Figure 5.1) and for the bispectrum the phase error will be approximately 0.1 radians (see Figure 5.4).

Assuming that the power spectrum and bispectrum are averaged over an observation of 100 s, if $t_0 = 10$ ms then incoherent averaging will reduce the noise levels by a factor of $\sqrt{100/0.03} \approx 58$ if the atmospheric errors are uncorrelated between exposures (which is a reasonable approximation providing that t_0 and r_0 are stable during the incoherent integration time). Thus, the atmospheric noise errors on the power spectrum will be at around the 1% level and on the closure phase at around the 0.1° level.

The systematic calibration errors in many cases are likely to be at least this large. Thus, atmospheric noise is not frequently an issue except in situations where extreme care has been taken to reduce calibration errors.

For faint objects, the atmospheric noise will be less than the detection noise. The faintest object which can be usefully observed will have an SNR after incoherent averaging of around 2 or so, meaning that if 1000 exposures are averaged the SNR per exposure is approximately 0.07. Thus, the detection noise in this case is a factor 20 larger than the atmospheric noise.

The conclusion that atmospheric noise is almost always negligible means that for most observations the dominant concerns when choosing parameters such as the exposure time are calibration accuracy and detection noise. The minimisation of the detection noise is the subject of the next chapter, and calibration issues are discussed in Section 8.8.

6

Interferometric observation of faint objects

The ability to observe faint objects is a key requirement for an astronomical instrument. Many types of object are intrinsically rare and therefore the closest examplars are far away and hence faint. For objects which are less rare, being able to 'go fainter' means that more exemplars of the class can be studied to give statistical validity to any findings.

The exposure times used in interferometry are typically much less than those used in other types of astronomical instruments – usually milliseconds instead of minutes or hours. As a result, a target which would be considered bright for many astronomical observations is considered faint in interferometric terms. This means that the majority of potential astronomical targets are likely to be in the faint-object regime for interferometry, because they will have been discovered using techniques which have intrinsically better faint-object sensitivity.

This chapter takes a quantitative look at the limitations of interferometry when observing faint objects. It looks at the trade-offs involved in adjusting the parameters of an interferometric observation such as the exposure time, with the aim of (a) determining the best parameter settings to use for observing faint objects and (b) determining the faintest possible object which can be observed under a given set of conditions.

Adaptive optics (AO) systems can be used to increase the signal-to-noise ratio (SNR) for an interferometric measurement, because they allow the use of a large-aperture telescope while ameliorating the negative effects of atmospheric wavefront errors on interferometric SNRs. In the same way, cophasing fringe trackers allow the use of long exposure times to improve the SNR of observations of faint objects.

Unfortunately, it is precisely on faint objects that the assumption that the atmospheric correction provided by active correction systems such as AO and fringe trackers breaks down. These systems need a sufficient number of photons from a reference object to accurately sense the wavefront perturbations, and the reference object is typically the object under study itself. If the object being studied does not provide enough photons then the level of correction will be worse, degrading the SNR of interferometric measurements, which were already low due to the faintness of the source.

This vicious cycle means that the limits to observing faint objects with interferometers often come from the limits of the active correction systems rather than the ability to build larger telescopes. Later sections explore these limits in more detail to understand the faint-object limits for interferometry. In the discussion that follows, the examples given will concentrate on the measurement of the power spectrum. The scaling of the SNR of the bispectrum will behave in a similar way.

6.1 The optimum exposure time

It was shown in Section 3.3 that changes of the atmospheric 'piston' phase with time cause fringe motion during a finite-length exposure and this causes the fringe pattern to smear out, with a resultant loss in fringe visibility. Thus, when observing bright objects, the exposure time should be reduced as far as possible in order to minimise the amount of calibration required of the visibility loss due to fringe smearing. When observing faint objects, the main concern is maximising the SNR, and this favours longer exposure times.

A fringe tracker can be used to mitigate the piston fluctuations but, as mentioned above, the assumption that perfect correction of atmospheric seeing is possible becomes less valid for faint sources. In the rest of this section it will be assumed that for faint sources a fringe tracker is a coherencing one, keeping the fringe envelope centered to within a few wavelengths but not having any effect on the fringe motion during an exposure. The assumption of a coherencing-only fringe-tracker operation at faint light levels is justified further in Section 6.4.

The optimum exposure time to use with a faint object needs to balance a number of competing effects. Longer exposures allow more photons to be collected in an exposure, and this will increase the SNR of the fringe measurements. At the same time, longer exposures are also subject to greater visibility losses due to fringe smearing, and this will tend to reduce the SNR of the fringe measurements. Finally, the shorter the exposure time, the greater the number of exposures that can be averaged in a given overall observing time and this will tend to increase the SNR of the incoherently averaged result.

To take all these effects into account, we need to find the exposure time τ_{exp} which maximises the SNR of an incoherently averaged observable such as the power spectrum or the bispectrum over some fixed observation time τ_{obs}. If an exposure time of $\tau_{exp} \ll \tau_{obs}$ is used then τ_{obs}/τ_{exp} exposures can be incoherently averaged and hence the averaged SNR is given by

$$\mathrm{SNR}_X(\tau_{obs}) = \mathrm{SNR}_X(\tau_{exp}) \sqrt{\tau_{obs}/\tau_{exp}} \qquad (6.1)$$

where $\mathrm{SNR}_X(\tau_{exp})$ is the SNR of the estimator X (for example the power spectrum P or the bispectrum T) for a single exposure of length τ_{exp}.

The scaling of $\mathrm{SNR}_X(\tau_{exp})$ with τ_{exp} will depend on whether photon noise or read noise is the dominant noise source. We will make the simplifying assumption that the exposures are in a low-light-level regime where $\mathrm{SNR}_X(\tau_{exp}) \ll 1$ for all the exposure times considered. This is clearly the regime in which optimising the exposure time is most important.

For photon-noise-limited measurements in the low-light-level regime the SNR of the power spectrum given in Equation (5.40) is

$$\mathrm{SNR}_P(\tau_{exp}) \propto \langle |\gamma(\tau_{exp})|^2 \rangle \overline{N}_{phot}(\tau_{exp}), \qquad (6.2)$$

where $|\gamma(\tau_{exp})|$ is the reduction of the visibility of the fringe due to fringe smearing over an exposure of length τ_{exp} and $\overline{N}_{phot}(\tau_{exp})$ is the mean number of photons received during the exposure. Thus, since $\overline{N}_{phot}(\tau_{exp}) \propto \tau_{exp}$, the SNR after incoherently integrating for a time τ_{obs} is

$$\mathrm{SNR}_p(\tau_{obs}) \propto \langle |\gamma(\tau_{exp})|^2 \rangle \sqrt{\tau_{exp}}. \qquad (6.3)$$

The results from Section 3.3 can be used together with Equation (6.3) to yield a graph of the variation in SNR_P with τ_{exp} as shown in Figure 6.1. It can be seen from this graph that, as expected, the SNR increases with exposure time for $\tau_{exp} \ll t_0$, but that fringe-visibility losses overwhelm the increase in the number of photons collected in the exposure for $\tau_{exp} > 1.6t_0$. Thus an exposure time of around $1.6t_0$ is optimal for low light levels and a fixed overall observation time τ_{obs}.

This exposure time will also be optimal in background-limited observations, such as in the mid-infrared where the dominant source of noise is noise from the background photons rather than photons from the source itself. In read-noise-limited situations, the variation of SNR with exposure time has a different functional form. The SNR of the power spectrum is given by

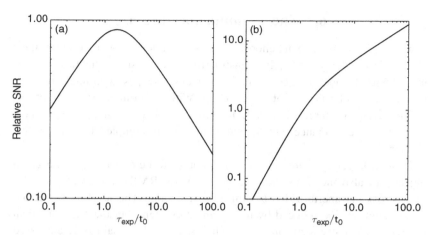

Figure 6.1 The SNR of the power spectrum as a function of integration time τ_{exp} for photon-noise-limited measurements (a) and read-noise-limited measurements (b) in the low-light-level regime. The SNR is normalised such that a system with a perfect fringe tracker that freezes the fringe motion would give an SNR of unity for an exposure time of t_0.

$$\text{SNR}_P(\tau_{exp}) \propto \langle |\gamma(\tau_{exp})|^2 \rangle \overline{N}_{phot}(\tau_{exp})^2, \qquad (6.4)$$

so the SNR after incoherent integration over a time τ_{obs} is

$$\text{SNR}_P(\tau_{obs}) \propto \langle |\gamma(\tau_{exp})|^2 \rangle \tau_{exp}^{3/2}. \qquad (6.5)$$

Figure 6.1(b) shows a plot of this function. It can be seen that in the read-noise-limited case the averaged SNR always rises as the exposure time is increased, albeit more slowly for exposure times greater than a few t_0. In practice this rise does not continue indefinitely because at some point the number of photons collected during an exposure becomes large enough that the exposure is photon-noise-limited and/or the assumption that $\text{SNR}_P(\tau_{exp}) \ll 1$ becomes invalid. It must also be borne in mind that using exposure times such that $\tau_{exp} \gg t_0$ means that the fringe visibility will be low due to fringe smearing, and this increases the danger that other effects such as the dynamic range of the detector, i. e. the ability to see small fluctuations in the intensity on a large background, may compromise the fringe measurement.

A similar analysis can be performed for the bispectrum showing that the exposure time which minimises the phase error in the photon-noise-limited case is about $2t_0$ (Buscher, 1988a).

6.2 The optimum aperture size

Few observers are able to choose what size of telescope to use for their observations. Nevertheless, it is instructive to understand the factors connected with the aperture size of the individual telescopes in an interferometer that allow the observation of faint objects. With a non-interferometric observation, using the largest available telescope is likely to guarantee the best faint-object performance, but the situation is more complex for an interferometer.

Using a larger aperture allows more photons to be collected from a given object, but also increases the root-mean-square (RMS) wavefront perturbation across the aperture due to atmospheric seeing. The level of wavefront perturbation can be reduced by use of an AO system, but there are practical limitations to the effectiveness of such a system. The main one, discussed below in Section 6.3, is that the AO system needs a bright reference source for wavefront sensing, and this bright source is usually the interferometric target itself. On fainter sources there is only enough light to sense the lowest-order modes of the atmospheric wavefront perturbations, so that the effective order of the AO system can be less than the maximum possible number of corrected modes that the AO system was designed for. In addition, AO systems are expensive and many interferometers only have tip–tilt correction available.

Thus, the use of larger apertures is usually associated with larger residual wavefront aberrations, which have the opposite effect on the SNR to the increase in flux collecting area. This is similar to the optimisation of the exposure time in the presence of temporal perturbations. There are, however, some differences. The first is that there is no incoherent averaging advantage to using smaller apertures as there is to using shorter integration times – this would correspond to using many parallel interferometers of smaller telescopes instead of a single interferometer with larger telescopes, and this is not usually an option within the budget constraints of most interferometric facilities.

An additional difference to the temporal optimisation problem is that spatial wavefront aberrations can have two types of effects on the SNR of the fringe measurements, depending on whether spatial filtering is used. If atmospherically perturbed beams are spatially filtered (using, for example, single-mode fibres), the fringe contrast can be increased at the expense of a compensating loss in the number of photons. To the author's knowledge, there is no temporal filtering equivalent (note that taking shorter exposures is not the temporal equivalent to spatial filtering but rather to using smaller apertures).

From Equation (5.37), the SNR of a power spectrum measurement for an aperture size D is given by

$$\text{SNR}_P(D) = \frac{\left\langle |F_{ij}(D)|^2 \right\rangle}{\sigma_P(D)} \tag{6.6}$$

where $\left\langle |F_{ij}(D)|^2 \right\rangle$ is the mean-squared coherent flux in a fringe pattern formed using aperture size D and $\sigma_P(D)$ is the RMS noise on the power spectrum for an observation with aperture size D. Equation (6.6) needs to be evaluated in at least six possible regimes depending on the type of hardware present and the wavelength range: two regimes corresponding to a spatially filtered and non-spatially-filtered beam combination combined with three regimes corresponding to photon-noise-limited detection, read-noise-limited detection and background-limited detection.

The variation of the numerator of the fraction in Equation (6.6) with D is similar when either a spatially-filtered combiner or a non-spatially-filtered combiner is used. In the case of a spatially-filtered combiner, the mean-squared coherent flux can be derived using Equation (3.60). The coherent flux will scale as

$$\left\langle |F_{ij}|^2 \right\rangle \propto \left\langle |\eta_i|^2(D) \right\rangle \left\langle |\eta_j|^2(D) \right\rangle D^2, \tag{6.7}$$

where $\left\langle |\eta_i|^2(D) \right\rangle$ and $\left\langle |\eta_j|^2(D) \right\rangle$ are the mean coupling efficiencies into a fibre from an aperture of diameter D and the factor of D^2 corresponds to the scaling of the total light collected with telescope area. The mean coupling efficiencies are given in Figure 3.17 and will typically be the same for both telescopes.

For a non-filtered beam combiner, the mean-squared coherent flux will be given by

$$\left\langle |F_{ij}|^2 \right\rangle \propto \left\langle |\gamma_{ij}|^2(D) \right\rangle D^2, \tag{6.8}$$

where $\left\langle |\gamma_{ij}|^2(D) \right\rangle$ is the mean-squared visibility loss for an aperture of diameter D and is plotted in Figure 3.14. By comparing Equations (6.7) and (6.8), and Figures 3.14 and 3.17, it can be seen that in both the filtered and unfiltered beam combiners the coherent flux will rise rapidly with diameter for small D but will rise less steeply or even fall after a 'cut-off' value of D/r_0, which depends on the order of correction.

The denominator in Equation (6.6) depends on which source of noise is dominant but can also depend on whether a filtered or unfiltered combiner is used. If the observation is photon-noise-limited, then in the low-light-level regime the noise is proportional to the mean number of photons per exposure

$$\sigma_P(D) \propto \overline{N}_{\text{phot}}(D). \tag{6.9}$$

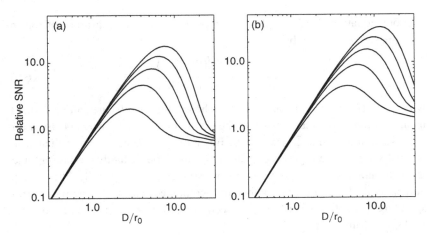

Figure 6.2 The SNR of photon-noise-limited fringe power spectrum measurements as a function of the interferometer aperture diameter D. The SNR is calculated for an interferometer using no spatial filter (a) or a single-mode spatial filter (b). Different lines show the effects of different radial orders of AO correction: the lowest line corresponds to the correction of Zernike wavefront modes up to radial order 1 (tip–tilt correction) and the uppermost line to a correction up to radial order 5. The SNR plotted is normalised to unity for a perfectly corrected aperture of diameter r_0, under the assumption that the source is sufficiently faint that the true SNR per exposure is much less than unity (the faint-source regime).

For an unfiltered beam combiner, the number of photons is given by

$$\overline{N}_{\text{phot}}(D) \propto D^2 \qquad (6.10)$$

whereas for a filtered combiner the coupling efficiency reduces the number of photons present:

$$\overline{N}_{\text{phot}}(D) \propto \left\langle |\eta(D)|^2 \right\rangle D^2. \qquad (6.11)$$

This means that the noise level is lower in the spatially filtered beam combiner compared to an unfiltered combiner, particularly for larger apertures where the coupling efficiency is poorer. This is borne out by the graphs of SNR versus diameter derived using Equations (6.6)–(6.11) and shown in Figure 6.2. It can be seen that in all cases there is an optimum diameter to use for a given level of AO correction. Above this diameter using a larger telescope actually decreases the SNR of the fringe measurements. In systems using spatial filtering, this optimum diameter is larger and the SNR at this optimum diameter is also larger, so if large telescopes are to be used, spatial filtering is of significant benefit.

At mid-infrared wavelengths and for the observation of very faint sources at shorter wavelengths, the dominant source of noise is the photon noise from thermal or other background radiation sources. The scaling of the noise from these sources with aperture diameter is different from the scaling for photon noise from the source. This is because, as the diameter of the telescope is increased, the increase in collecting area of the telescope is exactly compensated for by a corresponding decrease in the angular size of the diffraction-limited 'patch' on the sky from which radiation is received. In the case of a spatially filtered combiner the patch which couples into the spatial filter is almost exactly the same size as the diffraction limit of the telescope, whereas any well-designed but non-filtered combiner will use a focal plane 'cold stop', which is a few times the diffraction limit in size to reduce the amount of background radiation reaching the detector. This means that $\overline{N}_{\text{phot}}(D) \propto$ and hence $\sigma_P(D)$ is roughly independent of D.

A noise level that is independent of D is also seen when the fringe detection is read-noise-limited in the low-light-level regime. The resulting graphs of SNR versus D are shown in Figure 6.3. It can be seen that the behaviour in this case is quite different to when the main noise source is photons from the object itself. In the spatially filtered case, the SNR is maximised at a finite diameter as in the photon-noise-limited case. However, in the unfiltered case, the SNR can show a peak at some critical value of D, but going to even larger values of D causes the SNR to start to rise again, albeit more slowly, and thus the unfiltered system performs better at the very largest diameters.

The situation is slightly more complex in the background-limited case, as for very large values of D/r_0 the correction of the wavefront errors will be sufficiently poor that the majority of the flux will not be concentrated in a diffraction-limited core. Instead, the majority of the flux will be in a seeing-limited 'halo', and as a result the angular diameter of the cold stop needed to collect all the light from the source will need to be of order r_0/λ rather than a few times D/λ. This will increase the amount of background light allowed through the cold stop and means that for the largest apertures the SNR behaviour for the unfiltered combiner will be more like the source-photon-noise-limited case shown in Figure 6.2.

In all cases, the SNR benefit of using larger apertures is lower if the level of AO correction available is limited. As discussed in Section 6.3 this limitation is set not only by the expense and complexity of using higher-order AO systems but also by the light available for wavefront sensing, and this is particularly relevant for faint objects.

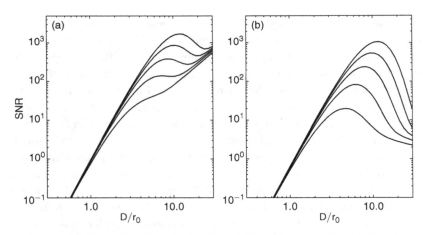

Figure 6.3 The SNR of read-noise-limited (a) or background-noise-limited (b) fringe power spectrum measurements as a function of the interferometer aperture diameter D. The SNR is calculated for an interferometer using no spatial filter (a) or a single-mode spatial filter (b). All other details are as in Figure 6.2.

6.3 AO on faint objects

The limitations of AO systems when observing faint targets arise from the limitations of wavefront sensing with faint reference sources. There are a number of different types of wavefront sensor used in AO, but for simplicity a single type of wavefront sensor will be used to illustrate these limitations. The Shack–Hartmann (or Hartmann–Shack sensor – there is some controversy about the naming) wavefront sensor consists of an array of lenslets placed at an image of the telescope pupil. These lenslets break the aperture of the telescope into a contiguous set of subapertures, and the light from each subaperture is focussed onto a different area of a detector, to form an array of spots as shown in Figure 6.4.

In the presence of an unabberated wavefront, all the spots will be on-axis for their respective lenslets, but in the presence of an aberrated wavefront the x and y location of the spots will be displaced by an amount which depends on the local tip and tilt across the subaperture sampled by the corresponding lenslet. These displacements therefore correspond to samples of the local wavefront gradient, and these gradient measurements can be integrated to form an estimate of the complete wavefront across the aperture. One way of performing this integration is to use the wavefront slope measurements to derive the coefficients of the Zernike modes and then to sum these modes together in the proportions given by the estimated coefficients.

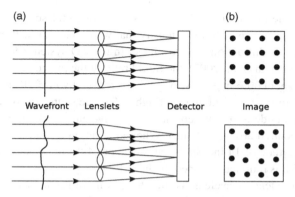

Figure 6.4 Schematic of a Shack–Hartmann wavefront sensor consisting of an array of lenslets (a) and the spot pattern seen on the detector (b). The upper row is for an unaberrated wavefront, which yields a regular spot pattern corresponding to the centres of the lenslets, while the lower row illustrates that the spots are displaced when the lenslets are illuminated with an aberrated beam. Diagram adapted from 'Adaptive optics tutorial at CTIO' by A. Tokovinin, http://www.ctio.noao.edu/~atokovin/tutorial/.

The larger the aperture to be corrected, the higher-order the AO correction needs to be in order to be effective (i. e. to reduce the residual wavefront phase aberrations to an acceptable level, e. g. of order a radian or less). A rule of thumb is that an aperture of size $D \gg r_0$ can be corrected using approximately $(D/r_0)^2$ degrees of freedom, so for example, a 4-m-diameter telescope would require the correction of about 1600 modes in order to give adequate correction when $r_0 = 10$ cm. There must be at least as many wavefront slope measurements as modes to be corrected and this leads to the result that the subapertures of the wavefront sensor need to be less than or of the order of r_0 in diameter.

The wavefront needs to be sampled before it has had a chance to change by of order a radian, and so the integration time needs to be less than about t_0. Thus the light available for each Shack–Hartmann spot is just the light which can be collected from the reference source over a patch of area approximately r_0^2 and during an integration time of t_0. Typically, about 100 photons are needed per spot in order to estimate the wavefront slope with adequate accuracy. As a result, a reference source of a certain minimum brightness is needed in order to give a certain minimum level of AO correction, and, importantly, this brightness is independent of the size of the telescope.

An example of this effect can be gleaned from the performance parameters of a typical AO system. A common measure of the performance of such a system is called the *Strehl ratio*. This measures the intensity of the light at the centre of the image of a point source seen through telescope and AO system, and compares this with the intensity that would be seen with perfect AO correction, i. e. a diffraction-limited image. In the AO system for the Keck II telescope (van Dam *et al.*, 2007), the Strehl ratio of the images at a wavelength of 2.2 μm is about 60% in good seeing (characterised by $r_0 = 20$ cm at a wavelength of 0.5 μm) when guiding on a bright reference source (a source brighter than 8th magnitude at the wavefront sensor wavelength, which is roughly in the R band centred around 600 nm). When guiding on a source which has an R-band magnitude of 14 the Strehl ratio is about 40% in the same conditions and at R=15 this ratio has dropped to about 15%.

At shorter science wavelengths the effect is more pronounced, in that the Strehl ratio typically begins to fall off even for much brighter reference sources. This is because r_0 and t_0 are smaller at shorter wavelengths so less light is available per subaperture and per exposure of the AO system.

6.3.1 Anisoplanatism

The reference source is often the science target itself, but if the science target is not bright enough then another reference source can in principle be used to drive the wavefront sensor. This reference source could be a star which is nearby in the sky to the science target. However, Figure 6.5 illustrates the problem with using an off-axis reference: the wavefront perturbations encountered by the light propagating from the reference source are not the same as those encountered by the science target and so the wavefront correction will not be perfect.

The larger the angle between the reference source and the larger the height of the turbulence layer which is causing the seeing, the larger the difference between the reference and target perturbations will be. This variation of the wavefront perturbations across the field is called *anisoplanatism* and the separation between two stars which experience a wavefront perturbation difference of 1 radian RMS is called the *isoplanatic angle*. It can be seen from the figure that this will occur when the separation of equivalent rays from the target and reference is of order r_0 at the height of the relevant layer of turbulence. Thus, the isoplanatic angle is given approximately by

$$\theta_{\text{isoplanatic}} \sim \frac{r_0}{h}, \tag{6.12}$$

where h is the height of the dominant turbulent layer. Taking a typical value for h of 5 km and for r_0 of 50 cm (corresponding to good seeing and a near-infrared science wavelength) then $\theta_{\text{isoplanatic}} \sim 10^{-4}$ radians or about 20 arcseconds.

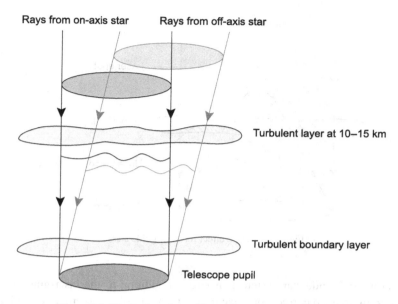

Figure 6.5 Illustration of the effects of angular anisoplanatism.

In order to provide acceptable correction, a bright star needs to be found within the *isoplanatic patch* around the target. The probability of such a star happening to be within the isoplanatic patch around a randomly selected target is about 50/50 if the isoplanatic patch is an arcminute in diameter and if reference stars with an R magnitude of 15 are considered 'bright'. However, both these conditions are rarely fulfilled and so for most interferometric targets, the best wavefront reference source is usually the science target itself.

6.3.2 Laser guide stars

An alternative to using off-axis reference stars is to use an artificially created reference star. This is the so-called *laser-guide-star* technology as shown in Figure 6.6, which relies on focussing a laser at some relatively high altitude in the atmosphere and using the light which is scattered back from the focus to sense the atmospheric wavefront perturbations.

Using a laser guide star overcomes the problem of finding a bright enough 'natural' guide star, but brings its own problems. The technology for laser guide stars is expensive and the wavefront correction provided by laser-guide-star techniques is imperfect in two ways. First, the lowest-order wavefront perturbations, tip–tilt and defocus, cannot be easily sensed using an LGS and

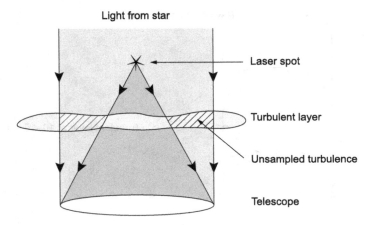

Figure 6.6 Cone effect for laser guide stars.

so an natural-guide-star system is needed in addition to the laser-guide-star system in order to correct for these terms. However, the low-order correction system does not need as many photons to provide the correction, and so fainter natural guide stars can be used.

A second problem of wavefront correction with laser guide stars arises because the guide 'star' is not at infinity. The wavefront perturbations seen on this reference will not be the same as those experienced by the target due to what is known as the *cone effect* or *focus anisoplanatism*. As shown in Figure 6.6, the backscattered light from the laser guide star samples a cone of the atmosphere, while the light from the science target samples a cylinder of the atmosphere. The differences between the turbulence sampled in these two volumes lead to differences in the wavefront perturbations, which grow worse as the telescope grows larger and causes the vertex angle of the cone to increase.

Currently, no interferometer is equipped with laser-guide-star AO on all the telescopes and so the majority of interferometric targets are observed using the target itself as the AO reference. When the reference is faint, the wavefront correction of large telescopes is less than perfect, and this provides a limit as to the faintness of the target which can be observed.

6.4 Fringe-tracking limits

Fringe trackers provide correction of temporal variations in the piston mode of the wavefront. Cophasing fringe trackers stabilise the fringes to allow the science fringes to be integrated for long periods but, just as with an AO system, the fringe tracker itself must have enough light to follow the fringes.

In the case of the fringe tracker, the wavefront sensor consists of a beam combiner which forms fringes and this fringe sensor unit will be prey to the same sorts of SNR issues when observing faint sources as any other interferometric instrument. One of these issues is sensitivity to spatial wavefront errors: any residual wavefront errors which are not corrected by the AO system cause either a loss of coupling efficiency of the starlight into the fringe tracker (for spatially filtered fringe trackers) or a loss in fringe visibility.

The fringe tracker must employ a minimum exposure time in order to measure the fringe motion. The shorter the exposure time on the fringe tracker, the less fringe motion there will be between exposures, but the less light is available to measure the fringe position and hence the greater the effects of noise on the fringe measurement. As a result there is an optimum exposure time for a given fringe-tracker light level, which balances the sensing noise against undersampling of the high-frequency fringe motion.

At a sufficiently low level, fringe tracking becomes ineffective because the fringe motion during the exposure time required to get sufficient SNR is too large. This light level is different for different types of fringe tracker. Two common types of fringe tracker are known as 'phase-tracking' and 'group-delay' fringe trackers, respectively, and these are discussed separately below.

6.4.1 Phase tracking

Measurements of the atmospheric optical-path-difference (OPD) perturbations in long-baseline interferometers such as those shown in Figure 3.4 show that the fringe phase excursions in any interferometer is likely to be many times 360°. Most cophasing fringe trackers rely on a technique known as 'phase unwrapping' or phase tracking to track such large excursions. In phase unwrapping, the fringe phase is sampled sufficiently often that the magnitude of the phase change between subsequent measurements due to atmospheric OPD fluctuations is never more than 180°. If this condition is achieved, then the atmospheric phase changes can be estimated by taking the difference between the fringe phases measured in consecutive exposures. The phase difference will be ambiguous to an integer multiple of 360° but the correct value of the phase difference can be inferred by choosing the value whose magnitude is less than 180°. By adding together these phase differences over multiple exposures, the phase change can be followed over arbitrarily many 360° cycles and so the science fringes can be stabilised.

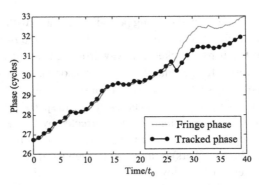

Figure 6.7 A sample fringe track showing a phase jump, which occurs when the fringe phase is changing rapidly compared to the sample time of the fringe sensor.

The maximum fringe-tracker exposure time allowable in order for phase unwrapping to work reliably is of order $t_0/2$ (Buscher, 1988b). If the exposure is longer than this then the phase shift between subsequent exposures occasionally exceeds $180°$ and an incorrect determination of the direction of fringe motion is made. If this happens the fringe phase estimated using the phase-unwrapping technique 'jumps' by $360°$ in a random direction as shown in Figure 6.7.

The fringe tracker will attempt to correct for the apparent phase jump by adjusting the internal delay by one fringe-tracker wavelength. If the science combiner is operating at a different wavelength, the science combiner will see a phase jump which is not a whole number of wavelengths. If the jump occurs during an exposure on the science combiner then the variation of the fringe phase will smear the fringe and cause a loss in fringe visibility.

A potentially more serious problem is that a series of phase-unwrapping errors will act as a random walk, tending to make the interferometer OPD wander away from the original tracked OPD value. Any polychromatic fringe pattern will have a fringe-visibility envelope (see Section 1.7), which decreases in modulus as the OPD moves away from the zero-group-delay location (see Section 1.8). Wandering away from the zero-group-delay location will therefore cause the modulus of the fringe visibility in the fringe tracker (as well as the science combiner) to decrease. Eventually, the SNR of the fringes in the fringe tracker will be so low that the fringe tracker will lose the fringes altogether and cease providing effective fringe tracking.

At low light levels, the SNR of the fringe-tracker phase measurements will be limited by detection noise. This noise will be significant because of the short exposure times used to avoid jumps in the phase unwrapping. At a sufficiently low light level, the noise in the phase measurements will be large enough to cause additional fringe-unwrapping errors, leading to similar consequences to those caused by using too long an exposure time. This occurs when the SNR of the fringe sensor measurements is less than about 2 for a single exposure (Buscher, 1988b). Sources which are fainter than this are problematic to fringe track using phase-unwrapping methods because fringe jumps are frequent.

Conversely, if the source being observed is brighter than this limiting magnitude, then the fringe tracking will likely to be accurate at the sub-radian level, as low phase noise is essential to avoid unwrapping errors.

6.4.2 Group-delay tracking

Phase tracking fails on faint sources as a result of the 360° ambiguity of the fringe phase, which means that an unwrapping error in a fringe-sensor exposure cannot be easily be recovered from in subsequent exposures. An alternative measurement that does not have this ambiguity is the position of the envelope of fringe visibility for a polychromatic fringe packet, as described in Section 1.7. In most circumstances the fringe envelope will have a peak at a single location in 'delay space' corresponding to the location of zero net group delay at the wavelength corresponding to the centre of the bandpass. If this location can be tracked then atmospheric disturbances of arbitrary magnitude can be followed, without requiring the disturbances to be tracked on timescales short enough so that the disturbances are much less than a wavelength.

The conceptually simplest way to track this envelope is to scan the delay lines rapidly backwards and forwards and to measure the fringe-visibility modulus as a function of delay as shown in Figure 6.8. The peak of the visibility envelope can then be found and the centre of the delay scan adjusted to keep the fringe envelope in the centre of the scan.

In order to be effective, the delay scan must be larger than the fringe envelope. As a result, much of the observation time is spent in regions of delay space where the fringe visibility, and hence the SNR of the fringe measurement, is low. An alternative technique for finding the fringe-envelope peak, which has better SNR in general, is known as *group-delay tracking*.

The group-delay method can be thought of as the inverse of a Fourier-transform spectrograph: instead of measuring the spectrum of an object by scanning a delay and observing a fringe envelope, the group delay uses spectrally dispersed measurements of fringes to reconstruct the fringe-visibility envelope as a function of delay. The implementation of the group-delay technique takes a number of forms, but a convenient way to visualise how it

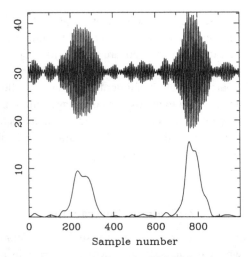

Figure 6.8 Fringe-envelope scans from the COAST interferometer. The change in OPD is approximately 60 μm during each 100-ms scan. The top trace shows the intensity as a function of time for two scans across the fringe envelope, and the bottom trace is the inferred envelope, which can be used to find the zero-OPD offset. From Thureau *et al.* (2003).

works is to consider a spectrally dispersed one-dimensional fringe pattern on a two-dimensional detector as shown in Figure 6.9(a).

The exact layout of the fringe pattern on the detector will depend on the design of the beam combiner, but in this example the x coordinate is a linear function of the delay difference τ between the beams interfering at that point on the detector and the y coordinate is proportional to the wavelength. The fringe pattern which would be seen on a target would be a sinusoidal pattern in the x direction at each wavelength whose spatial frequency is proportional to wavelength. For an unresolved source at the phase centre and with no net instrumental or atmospheric delay errors the peaks of all the sinusoids are aligned with each other at the centre of the fringe pattern, as shown in Figure 6.9.

If the fringe pattern intensities are remapped onto a new set of coordinates ($\phi \propto x/\lambda, \nu \propto 1/\lambda$) then all the horizontal fringe patterns will have the same spatial frequency and the fringe pattern with no delay errors will appear as a vertical sinusoidal pattern as shown. If an instrumental delay error τ_{ij} is introduced on baseline ij then the fringe patterns will shift in phase by an amount which depends on wavelength as

$$\phi(\nu) = \tau_{ij}\nu. \tag{6.13}$$

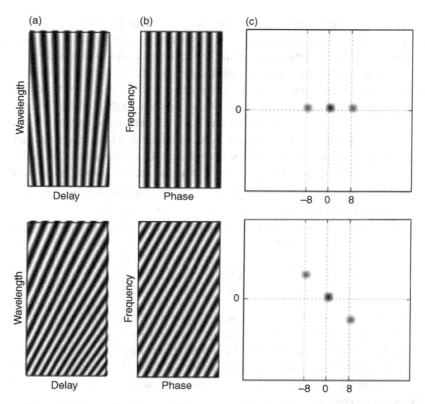

Figure 6.9 Spectrally dispersed fringe patterns (a), the fringe patterns remapped so that the fringe crests are parallel (b) and the power spectra of the remapped fringes (c). The upper row corresponds to zero OPD while the lower row corresponds to a finite OPD, which causes the fringe phase to change linearly as a function of wavelength.

This linear phase change with frequency v appears as a change in the slope of the fringes as shown in the lower row of Figure 6.9. Thus, by determining the slope of the fringe phase with frequency the delay offset can be deduced.

The slope of the fringe crests is easy to identify 'by eye' at high light levels from the intensity patterns in Figure 6.9. At low light levels the slope can be determined by taking a two-dimensional Fourier transform of this pattern. The magnitude of the Fourier transform will have a peak at the origin corresponding to the 'DC' level of the pattern and a symmetric pair of peaks corresponding to the sinusoidal intensity modulation. As the slope of the fringes changes, the peak will move along a line with constant spatial frequency in the x direction. Taking a slice of the Fourier transform along this line will show a peak in magnitude whose position depends on the delay error τ_{ij}. The group-delay estimate is then an estimate of τ_{ij} based on the estimated position of the peak.

This group-delay method uses all the flux received at all wavelengths simultaneously and so works well at low light levels. An alternative way of viewing the group-delay estimation process is one determining a best-fit value of the delay τ_{ij} as a two-step process (Basden and Buscher, 2005). In the first step (corresponding to Fourier transforming in the x dimension and taking the Fourier component corresponding to the fringe frequency) the fringes at each wavelength are analysed to yield a set of coherent flux estimates for all wavelength channels $\{\hat{F}_{ij}(\nu_k), k = 1, \ldots, N_{chan}\}$. In the second step a set of trial delays $\{\tau_{ij,m}, m = 1, \ldots, N_{delay}\}$ is generated, and for each trial delay the coherent flux estimates are 'phase rotated' to compensate for the assumed delay and then added together to give the coherent flux of the 'synthetic white-light fringe' at that delay, given by

$$F_{ij,m} = \sum_{k=1}^{N_{chan}} F_{ij}(\nu_k)e^{2\pi i \nu_k \tau_{ij,m}}. \qquad (6.14)$$

This latter step corresponds to a Fourier transform over the y dimension if the channels are sampled evenly in frequency. In the absence of noise and for an unresolved source the amplitude of the white-light fringe $|F_{ij,m}|$ is maximised when the trial delay equals the true delay. This is equivalent to finding the trial delay which best 'unwraps' the linear change of phase with frequency caused by the delay error τ_{ij}.

The precision with which the position of the peak in delay space can be determined is related to the SNR of the data and the width of the group-delay peak. The width of the peak is given approximately by

$$\Delta\tau \sim (\Delta\nu)^{-1}, \qquad (6.15)$$

where $\Delta\nu = \nu_{max} - \nu_{min}$ is the total bandwidth of all the wavelength channels combined.

This result can be derived by considering Equation (6.14) as Fourier transform of the coherent fluxes over an infinite bandwidth multiplied by a 'top hat' of width $\Delta\nu$. Comparing this to the results in Section 1.7, it can be seen that the width of the peak is also the width of the fringe envelope of the synthetic white-light fringe, i.e. the fringe formed by summing pixels along the wavelength direction in the dispersed fringe pattern.

If $\Delta\nu \sim \nu_0$, where ν_0 is the central frequency of the spectral channels, the width of the peak is comparable to a wavelength and so precisions in the delay estimate of this order are possible. In order to get higher precision, the phase

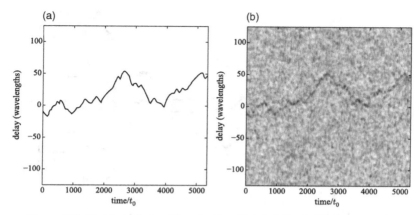

Figure 6.10 Simulated group delay signal at high light levels (a) and at a light level corresponding to the lowest SNR at which the fringe envelope can be reliably tracked (b). Each column of pixels corresponds to the time-averaged group-delay fringe power for different trial delays at a given instant in time.

of the synthetic white-light fringe phasor given in Equation (6.14) can be used to follow the sub-wavelength motion of the fringe, in a similar way to that used in phase tracking.

At lower light levels, group-delay tracking is generally used for coherencing and not cophasing.

The accuracy requirements for coherencing are lower and so the signal can be averaged for longer. In order to make peak identification more reliable the power spectrum as a function of trial delay can be integrated incoherently over many exposures in order to increase the SNR.

Simulations of group-delay tracking (Basden and Buscher, 2005) show that the group-delay power spectrum can be incoherently integrated for 20–$50 t_0$. Figure 6.10 shows that the group-delay signal can be reliably extracted when $\text{SNR}(\hat{F}_{ij}) \gtrsim 0.6$.

In Section 6.4.1 it was stated that phase tracking is possible when $\text{SNR}(\hat{F}_{ij}) \gtrsim 2$. Taking into account the different exposure times ($0.5 t_0$ for phase tracking and $1.6 t_0$ for group-delay tracking), the corresponding different visibilities and the scaling of $\text{SNR}(\hat{F}_{ij})$ with the number of photons in photon-noise-limited observations, this implies that group-delay tracking can be used on objects more than ten times fainter than the objects which are at the limit for phase tracking.

As a result of being able to work on targets which are considerably fainter, group-delay tracking can be used on a far wider range of targets than phase tracking. The fact that group-delay tracking only does coherencing and not

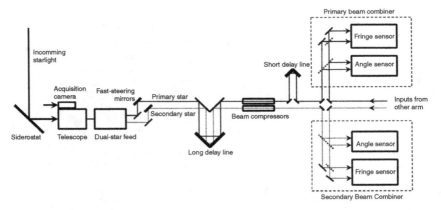

Figure 6.11 Schematic design of the dual-star system in the PTI. Two parallel beam combiners measure fringes simultaneously on a primary and secondary star. From Colavita *et al.* (1999).

cophasing at low light levels implies that the majority of faint targets will be observed using multiple short exposures on the science camera rather than relying on a cophasing fringe tracker to allow long coherent integrations.

6.4.3 Off-axis fringe tracking

The fact that fringe tracking only works on relatively bright targets restricts the range of targets which can be observed. A potential way around this restriction is to use off-axis stars as references for the fringe tracker, as is done in AO. Systems which accomplish this effect are known as 'dual-star' or 'dual-feed' systems.

A conceptual outline of a dual-star system is shown in Figure 6.11. The light from two different stars with angular separations of many arcseconds are split by a star separator at the focal planes of the unit telescopes and sent down parallel paths to two separate beam combiners. The brighter of the two stars is used to sense the atmospheric piston variations and these are used to provide corrections to a delay line which is common to the optical paths of both beam combiners, thereby stabilising the fringes in both combiners.

Dual-feed systems incorporate differential delay lines, which remove the additional geometrical delay difference between the two combiners due to the fact that the two stars are at substantially different locations in the sky

The dual-feed system implemented in the PTI was not primarily intended to allow the scientific observation of faint targets. The main reason for this is that, in order to provide good cophasing of a faint target, there needs to be a bright enough reference star within the isoplanatic patch of the science target to drive the fringe tracker. Fringe trackers typically require brighter references in order to operate than AO systems, and so the probability of finding bright-enough fringe-tracking references is even lower than for finding wavefront references for an AO system, perhaps less than 1% in a typical interferometer.

Instead, dual-star systems are typically aimed at astrometric science targets on bright nearby stars. These science targets are used to drive the fringe tracking, and the second star is a faint star which is used as an astrometric reference. The cophasing provided by the target star serves to allow the SNR of the faint reference fringes to be increased by using long integrations. Faint reference stars are plentiful and therefore are not a problem to find within the isoplanatic patch.

6.4.4 Laser guide stars for fringe tracking

There is currently no equivalent of a laser guide star for single-telescope AO which can be used to provide an artificial signal for fringe tracking in interferometry. There are a number of reasons for this. One of these is the difficulty of providing a sufficiently small laser spot (milliarcseconds in size) so that the fringes from the spot have high contrast.

A second and more serious problem is that the 'cone effect' is difficult to overcome for long-baseline interferometers because the severity of the cone effect becomes larger, the larger the ratio between the baseline and the height of the laser spot becomes. Techniques for overcoming the cone effect for large telescopes, such as the use of multiple laser guide stars, rely on being able to 'map out' the turbulence as a function of altitude across the whole aperture, and this is difficult to do when the aperture consists of isolated patches as it does in an interferometer.

In the near term, the best hope for the use of laser guide stars in interferometry is to correct the non-piston atmospheric aberrations across each of the apertures of the array. If large-enough apertures can be corrected in this way, this will allow sufficient light to use fainter targets as natural references for fringe trackers aimed at stabilising the piston mode.

6.4.5 Bootstrapping

The previous sections show how important the SNR of fringe measurements are for successful fringe tracking and therefore for successful observations of a given target. This SNR depends on both the number of photons per exposure and on the fringe visibility. The latter is affected by both instrumental

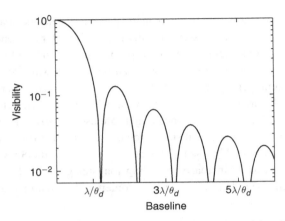

Figure 6.12 The object visibility modulus as a function of baseline when observing a uniform disc.

effects and by the visibility function of the object being observed. The object visibility function is a generally decreasing function of baseline length, so the SNR for fringe tracking will also decrease with baseline length. This can cause problems for fringe tracking on long baselines.

As an illustration of this problem we can consider the observation of features on the surface of a star, for example convection cells on the surface of a supergiant star such as Betelgeuse. To first order, the stellar surface can be modelled as a uniform disc of angular diameter θ_d, and so the fringe visibility as a function of baseline length will be an Airy function, depicted on a log scale in Figure 6.12.

In order to be sensitive to structures on the disc surface which are ten times smaller than the disc (in other words to be able to make a 10×10 resel image of the disc), the spatial frequency sampled by the longest baseline used must have at least five peaks and five troughs across the stellar disc. This corresponds to a longest baseline, which is at least $5\lambda/\theta_d$ in length. From Figure 6.12 it can be seen that the object visibility function in this region has a maximum of about 2%.

In the photon-noise-limited regime, the SNR is a monotonic function of $V \sqrt{N_{\text{phot}}}$ and so the 50-fold reduction in fringe contrast compared to an unresolved source has the same effect on the SNR as a reduction in the photon rate by a factor of about 2500. In other words, over-resolving the stellar disc by this factor is equivalent in SNR terms to exchanging 8-m diameter telescopes for 16-cm telescopes. As a result, fringe tracking on such objects may fail on the long baselines when it works adequately on shorter baselines.

The net result is that it may not be possible to observe objects on the long baselines where the object shows interesting structure. What counts as a 'long' baseline for these purposes depends entirely on the size of the object under study. In the case of Betelgeuse with an angular diameter of $\theta_d = 40$ mas, $5\lambda/\theta_d = 12.9$ m and this is comparable to the *shortest* baselines available on many interferometers.

The situation of not being able to track fringes on baselines needed to resolve interesting structure is more common in optical interferometry than in radio interferometry. The main reason is that dependence of the SNR on the visibility is weaker in the background-noise-limited regime, which is prevalent at radio wavelengths, than in the photon-noise-limited regime, which can be encountered at optical wavelengths.

A secondary reason is due to the typical spatial structure of the targets at the two wavelengths. Many targets at radio wavelengths, for example radio galaxies, contain a bright unresolved 'core' surrounded by a much larger region containing interesting structure. The presence of the core means that the visibilities remain high even on baselines which are long enough to resolve the rest of the structure in the object. At optical wavelengths, many of the sources being observed have a structure where majority of the emission comes from an object, such as a stellar surface, which is on the same scale as the interesting features. Therefore, when the baseline is long enough to see detail, the 'core' is also resolved – such objects can be termed 'resolved-core' objects. Nevertheless, there are many objects which have a brightness structure at optical wavelengths that are like the 'compact-core' radio targets, for example extended discs around stars, and for such objects fringe tracking on long baselines is less of an issue.

There are a number of ways around the fringe-tracking limitations which occur when observing resolved-core objects. One technique, known as *wavelength bootstrapping*, is to track fringes at a different wavelength from the science wavelength, where the fringe visibility is higher on the same baseline.

If the object has roughly the same apparent size at all wavelengths, then the fringe visibility observed on a given baseline will tend to be greater at longer wavelengths than at shorter wavelengths. For example, using a fringe-tracking wavelength of 2 μm and a science wavelength of 500 nm would be advantageous, as the (u, v) coordinate sampled by the fringe tracker would be four times smaller than the (u, v) coordinate sampled by the science instrument. Thus, the fringe tracker could be inside the first lobe of the visibility function of a uniform-disc object such as Betelgeuse while the science instrument is sampling the fourth lobe at much lower visibilities.

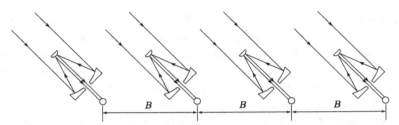

Figure 6.13 A set of telescopes arranged in a baseline-bootstrapping configuration.

Alternatively, the fringe tracker can be used at wavelengths where the apparent object size is much smaller and so unresolved. For example, when observing a star with a disc of warm dust around it, the disc may only be visible at mid-infrared wavelengths. A fringe tracker operating at a 2-μm wavelength might only see the star itself and would have high-visibility fringes if the star is unresolved. At the same time, a mid-infrared science instrument operating at 10 μm would see the disc emission and hence potentially resolve structure in the disc.

In both cases the science beam combiner would see quite low visibilities and hence the SNR in a short exposure would be low. However, because the fringe tracker is seeing higher visibilities it is able to provide effective fringe tracking, and the fringe measurements on the science camera can be integrated (coherently if the fringe tracker is a cophasing tracker and incoherently if not) over long periods to provide useful SNR data.

Another technique for allowing fringe tracking on resolved-core sources is called *baseline bootstrapping*. This requires an array of telescopes arranged in a 'chain' of short baselines making up a longer baseline such as that shown in Figure 6.13. Fringes can be measured on all the short baselines because the source is unresolved on these baselines and so the fringe visibilities are high. The fringe motions on the short baselines can be used to drive piston actuators to compensate for the piston phase differences between adjacent telescopes. As a result of this, the phase differences between all pairs of telescopes in the chain will also be compensated, and so fringe tracking on the shortest baselines provides fringe tracking on the longest baselines as a natural byproduct.

Baseline bootstrapping is most effective if the nearest-neighbour telescopes are as close together as possible to allow the highest possible visibilities for fringe tracking while keeping the longest baselines as long as possible, in order to resolve interesting structure in the target. These competing factors

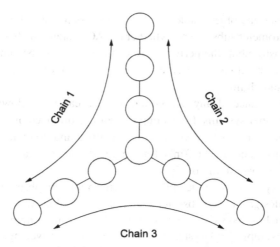

Figure 6.14 The baseline bootstrapping layout of the MROI.

tend to favour an array of equi-spaced telescopes, a 'redundant' arrangement which is less than optimal for covering a wide diversity of (u, v) plane
spacings.

For a linear chain of telescopes, the ratio of the longest to the shortest
baseline increases linearly with the number of the telescopes, whereas in a
non-redundant array this ratio can increase approximately as the square of the
number of telescopes. This is less of a problem in two-dimensional arrays of
telescopes, as even with an optimal non-redundant design, which covers the
(u, v) plane evenly, the ratio of the longest to the shortest spacing can at best
grow linearly. Figure 6.14 shows a bootstrapping layout used in the MROI.
The ten telescopes are arranged in an equispaced Y-shaped configuration, so
the array can be viewed as a set of three 'bootstrapping chains' each consisting
of seven telescopes, oriented at 120° angles to one another. The (u, v) coverage
of the array is shown in Figure 2.6; only 20% of the baselines are redundant in
this case.

6.5 Faint-object limits for interferometry

The above analysis has confirmed that atmospheric phase perturbations serve
to reduce the signal-to-noise ratio of interferometric measurements on faint
objects. The effects of the atmospheric perturbations can be ameliorated using
active systems such as AO and fringe tracking, but these themselves do not

work well when the object under study is faint. As a result, for each of these major interferometer subsystems, whether the AO system, the fringe tracker, or the science instrument, the performance of the subsystem becomes unacceptable when the object being observed is below some limiting faintness, known as the magnitude limit.

If the performance of any one system falls, it can bring down the performance of the other systems. For example, poor performance of the AO system will lead to large residual wavefront perturbations and this will in turn bring down the performance of the fringe tracker and the science camera, potentially making the observation impossible.

Which subsystem limits the ability to perform a given observation depends both on the design of the subsystems and also the colour of the object being observed. Each subsystem typically takes light from different wavelength ranges; for example, AO systems often take visible-wavelength light while fringe trackers and science instruments might both be at infrared wavelengths. Many interferometric targets are quite 'red' in that they are significantly brighter at infrared wavelengths than at visible wavelengths, and so it is possible for the AO system to perform poorly on objects which are bright at the science wavelength.

The performance of all these subsystems depends on the number of photons received from the object during the coherence time of the atmospheric fluctuations t_0 and over a coherence patch of order r_0^2 in size. Thus, the magnitude limits strongly depend on the prevalent seeing, and so siting interferometers in places with intrinsically good seeing and scheduling observations of faint sources during the periods of best seeing can dramatically improve the magnitude limits.

Since r_0 and t_0 are wavelength-dependent, the magnitude limits are typically better at longer wavelengths. At wavelengths longer than about 2 μm, the background noise due to photons emitted by the atmosphere and the instrument becomes the dominant noise source and the noise increases rapidly with increasing wavelength. As a result, the sensitivity of interferometers tends to peak at near-infrared wavelengths, which have the best combination of large r_0 and t_0 and low backgrounds.

At the time of writing of this book, the best reported magnitude limits for interferometric observations were visible magnitudes of around $m = 7$ (Mourard et al., 2009) and infrared magnitudes of around $m = 11$ (Petrov et al., 2012). Of note is the fact that both of the faintest limiting-magnitude measurements reported above were made using the group-delay fringe-tracking method with the science instrument itself acting as the fringe tracker.

Improvements in limiting magnitude are likely to come from improvements in detectors and in the performance of the interferometer 'infrastructure', such as reducing vibrations and increasing throughput and these improvements are likely to result in targets some 2–3 magnitudes fainter being observable. Further improvements will likely need the use of laser guide stars or going into space.

7

Observation planning

Interferometric observations need to be planned in advance, because observing time on an interferometer is a scarce resource. This planning is often carried out in the context of a competitive proposal scheme like that operated by most research telescopes. The typical process involved is that potential observers submit proposals to a 'time-allocation committee' or similar body, this body ranks the proposals and the most highly ranked proposals are given time.

The criteria for ranking the proposals can be somewhat subjective, but usually involve a combination of the technical feasibility of the proposed observations and the likely scientific value of the information that will result from them. In order to be highly ranked, proposals must address both of these aspects, so developing such a proposal requires expertise in two areas: familiarity with a relevant science field to understand where the gaps in knowledge are and an understanding of the ways in which different kinds of interferometric observation can provide critical information to fill these gaps.

This chapter looks at the interaction between the scientific and technical aspects of an interferometric observing proposal, with the aim of highlighting the most important areas to consider. The details of writing a competitive astronomy proposal are beyond the scope of this book, but an online search for 'how to write an observing proposal' turns up many useful links on the topic.

7.1 Example proposal

The material in this chapter will be illustrated in part through an example proposal for the VLTI, which was awarded observing time in August and September 2012. The science background to the proposal is that of Mira variables, named after the prototype star of the class otherwise known as Omicron Ceti. The name Mira is latin for 'wonderful' or 'astonishing', a reference to its regular appearance and disappearance from the visible sky; pulsations in the star cause large changes in brightness on periods of a few hundred days.

Many Miras exhibit hydrogen emission lines in their spectra, thought to be signatures of shocks that accompany the pulsation of the star. The passage of these shocks may influence many of the physical conditions in the extended atmosphere of Miras but details of the locations and geometries of these shocks are poorly constrained. Interferometry has the required angular resolution to image these shocks and so could potentially cast a light on shocks and their roles in Mira atmospheres. The details of what kind of observations would best achieve this aim will be discussed in later sections.

7.2 Target selection

The subject of any astronomical observation is a selected object or class of objects that are of scientific interest for some reason. To be a suitable target for an interferometric observation the objects must pass a few additional tests.

7.2.1 Angular size

The angular scales of the objects, and especially the features of the objects which are of interest, need to be in an appropriate range. It is clear that targets which are known to be too small to be resolved by any available interferometer will be of little interest, as in these cases the visibility information can be deduced without needing to use any interferometer time: the visibility will be unity on all observable baselines.

The angular resolution of an interferometer is given approximately by

$$\Delta\theta_{res} \sim \frac{\lambda}{B_{max}} \text{ radians} \approx 200\frac{\lambda/\mu m}{B_{max}/m} \text{milliarcseconds}, \tag{7.1}$$

where B_{max} is the maximum baseline in the interferometer. Current interferometers have maximum baselines in the 200–400-m range so targets typically need to be bigger than about 0.2 milliarcseconds to be viable candidates for interferometric observation.

Less obvious is the fact that targets which are too large are also not good candidates for interferometric observation. One reason is that the structure on scales large enough to be accessible on single telescopes is typically easier to image using these systems than with an interferometer.

There will always be targets which are large enough for the target as a whole to be resolved by single telescopes but having structure on smaller scales appropriate to interferometic observations. In such cases there still remain a number of problems.

One of these is related to the fact that the source visibility for any given object typically falls off with baseline. From the examples in Section 2.2 it can be seen that the characteristic baseline at which the visibility falls by half is given approximately by

$$B_{\text{half}} \sim 200 \frac{\lambda/\mu\text{m})}{\theta/(\text{milliarcseconds})}\text{m} \qquad (7.2)$$

for an object of angular size θ.

The shortest baselines available on most interferometers are greater than 2 m. Thus, for observing wavelengths of order 1 μm and targets larger than about 100 milliarcseconds in size, the visibility will be low on all baselines observed. This low visibility has two important consequences: first, the signal-to-noise ratio (SNR) of the visibility measurements will be low, especially if the source is faint, which means that fringe tracking may not be possible and so the object may not be observable. Secondly, image reconstruction will be compromised because the region of the Fourier plane which has the most signal, i. e. the high regions of the source visibility function that occur predominantly at low frequencies, is not sampled. This low-frequency flux is said to be 'resolved out' as it is essentially 'invisible' to the interferometer and this missing flux can cause problems in image reconstruction, especially if it accounts for the majority of the flux in the image.

A related issue for large objects is the interferometric field-of-view discussed in Section 2.4.3 and given by

$$\theta_{\text{FOV}} \sim 200 \frac{\lambda/\mu\text{m}}{\Delta B/\text{m}}\text{milliarcseconds,} \qquad (7.3)$$

where ΔB is a measure of the 'typical' gaps in baseline sampling. There will usually be a gap in the sampling at low frequencies due to the issues of telescope spacing mentioned above, and for similar reasons there will typically be gaps in Fourier coverage at high frequencies of similar size. If these gaps are of order 2 m or more then observing objects which subtend angles larger than about 100 milliarcseconds at wavelengths of order 1 μm could result in ambiguities in the image because of Fourier undersampling.

7.2.2 Brightness

Chapter 6 explains why interferometers have difficulties observing objects fainter than a given limiting magnitude. These magnitude limits come in three forms: limits to the adaptive optics (AO) systems on each telescope; limits to the fringe acquisition and tracking; and limits to the SNR of the science data. Each of these limits relates not only to the total flux from the object but also to the angular size of the region from which this flux emerges: if this region is too big, then the object needs to be correspondingly brighter.

In the case of the AO system, the magnitude limit relates to the flux from a region compact enough to allow the wavefront sensor to work properly. Typically this size is of order an arcsecond or so across. If the target is diffuse so that it has a significant fraction of its flux on scales larger than this then usually it is better to use an off-axis star as a wavefront reference, as it is often the case that the wavefront sensor simply does not work on diffuse objects, no matter how bright they are. Particularly troublesome are binary stars with separations of a few arcseconds as the wavefront sensor may randomly switch from tracking one star to tracking the other or track at a point in between the two.

7.2.3 SNR estimation

In the case of the fringe tracking and science data, the flux limits relate to the SNR of the fringe data. If the SNR is too low in the fringe tracker then the zero-OPD position cannot be found and/or tracked and so no useful science data can be taken. If the fringes can be found in the fringe tracker but the SNR is too low in the science instrument, then the science data will be of limited value.

An acceptable value for the SNR is dependent on the application. For fringe trackers the SNR of the fringe-tracking signal, be it the fringe phase for cophasing or the group delay for coherencing, has to be greater than about 2 for tracking to work reliably. This SNR needs to be reached after an averaging time that is less than the time it takes for the quantity being tracked to change by a significant amount; this can be a few milliseconds or a few seconds depending on the type of tracking desired.

For the science instrument the acceptable SNR value depends on what precision of constraint is needed for the given science. At the low end, there are relatively few applications which can make use of data where the *averaged* SNR is less than about 2 or so; typical requirements are SNRs of 5–10 at the minimum while for detecting low-contrast objects, such as faint companions to bright objects, the SNR may need to be hundreds or even millions.

The SNR *in a single exposure* can be less than unity if many exposures are averaged. For example, with a 10-ms exposure time the SNR of the incoherently averaged data will be 100 times larger than the per-exposure SNR if

100 s of data are averaged. As a result per-exposure SNRs of 0.1 are acceptable because they can result in final SNRs of 10.

Several formulae allowing the calculation of SNRs for different observables such as the power spectrum and bispectrum are given in Chapter 5. To use these formulae often requires knowledge of the seeing, the interferometer throughput, the read noise of the detector and many other factors. Luckily, for most instruments the SNR has been calculated for a reference case (typically a limitingly faint source of a defined type), and so determining whether a given observation can be achieved at an acceptable accuracy requires only an understanding of the *scaling* of the SNR with different parameters.

The formulae will scale differently for different noise regimes, for example atmospheric-noise-dominated versus read-noise-dominated versus photon-noise-dominated. For bright objects, the SNR is atmospheric-noise-limited and independent of the source flux. For fainter objects, the limiting source of noise is detection noise of one form or another. In this case the SNR depends not only on the flux from the object but also on the fringe visibility. The fringe visibility is proportional to the source visibility function on the baselines being observed so flux limits are often quoted for an unresolved source, where the source visibility is unity on all baselines. As most sources of interest are resolved, the magnitude limit needs to be reduced by an amount depending on the predicted source visibilities.

For read-noise-limited or background-noise-limited observations, Equation (5.28) shows that the SNR of the fringes on a given baseline is proportional to the coherent flux. For some interferometric instruments, the limiting magnitude is therefore quoted as a coherent flux or 'correlated flux', computed by multiplying the source flux by the source visibility on the baseline in question. The correlated flux is often quoted as a 'correlated magnitude':

$$m_{\mathrm{corr}} = m - 2.5 \log_{10}(V), \tag{7.4}$$

where m is the magnitude of the object in some waveband and V is the object visibility in that waveband.

7.2.4 Surface brightness

An important concept relating the ideas of the previous two subsections is that of *surface brightness* (known in optics as *spectral radiance*). The apparent brightness or flux of an object is the power received from the object per unit collecting area per unit of frequency bandpass. The surface brightness of a particular region of an object is the flux received from that region divided by the solid angle subtended by the region at the location of the observer. Conversely, the flux is the surface brightness integrated over the total 'angular area' of the object: the flux is what is usually measured by a telescope that is unable to resolve a given object. Astronomers typically measure flux in janskys (Jy), where $1\,\text{Jy} = 10^{-26}\,\text{W/m}^2/\text{Hz}$, and so surface brightness can be measured in Jy per steradian (Sr).

Importantly, the surface brightness of a given region does not change as the object is moved away from the observer (providing the intervening medium is transparent), because the solid angle subtended by the region decreases at exactly the same rate as the flux received from the region decreases. Thus, if the surface brightness of an object is known, the angular extent of an object can be estimated from the measured flux from the object without knowing the distance to the object or its physical size.

If the object is a star or other object which can be modelled as a black-body emitter, then the surface brightness of the object can be estimated if there is information about the temperature of the emitter. The surface brightness at frequency v of a black body at temperature T is given by the Planck radiation law:

$$B_v(T) = \frac{2hv^3}{c^2} \frac{1}{e^{hv/(k_BT)} - 1}.$$ (7.5)

Thus, knowing the flux and the temperature of a target allows the angular size to be estimated and hence the range of appropriate baselines to be determined.

This technique can be used, for example, in the planning of observations of M-type giants and supergiants. These are favourite targets for interferometric observers partly because they are bright and have relatively large angular sizes. To observe these at a wavelength of 1.6 μm using an interferometer such as CHARA whose minimum baseline is around 30 m and maximum baseline is about 330 m implies selecting targets with angular sizes of between about 1 and 11 milliarcseconds. These supergiants have typical temperatures of about 3500 K, and so Equation (7.5) predicts a surface brightness of $8 \times 10^{-9}\,\text{W m}^{-2}\,\text{Hz}^{-1}\,\text{Sr}^{-1}$ at a wavelength of 1.6 μm.

The solid angle subtended by a disk of angular diameter $\Delta\theta \ll 1$ radian is approximately $\pi\Delta\theta^2/4$ steradians, and so stars of diameter 1 and 11 milliarcseconds will have fluxes of about 15 and 1800 Jy, respectively. The flux of a zero-magnitude star is about 1020 Jy in the H-band (Bessell *et al.*, 1998) and so these fluxes correspond to H-band apparent magnitudes of −0.6 and 4.6, respectively. There are dozens of M giants with H-band magnitudes brighter than −0.6; these need to be excluded from the target list because they are too large for observation with these baselines.

7.2.5 Mira example

Mira variables are even cooler than M giants and supergiants and so for a given brightness have even larger angular sizes. The prototype of the class known simply as Mira (o Ceti), is amongst the brightest and has an angular diameter of around 30 milliarcseconds. Putting this into Equation (7.2) shows that baselines of less than 13 m are required to adequately sample the low-frequency portion of the visibility function at a wavelength of 2 μm. The shortest VLTI baselines available at the period of the proposal were around 11 m and so were adequate.

In the chosen instrumental configuration (the AMBER instrument on the VLTI in HR–K mode, see below), the correlated magnitude limit was quoted as $K_{corr} = 5.5$. The Mira prototype has a K-band magnitude of −2.5; since Mira is resolved on the chosen baselines, a typical fringe visibility can be assumed to be of order 0.1 and so from Equation (7.4) its correlated magnitude is $K_{corr} = 0$, well inside the limit.

In order to achieve the quoted magnitude limit on the dispersed fringes the VLTI fringe tracker FINITO is required to allow long exposures (typically a second or more) on AMBER. FINITO has difficulty tracking on sources with H-band visibilities of less than about 5%, and so with a resolved source like Mira it is possible that the tracker would not work.

An alternative would be to record fringe data without the tracker using short exposures. No magnitude limit for this mode is given in the official documentation, but one can be estimated from the $K_{corr} = 6$ magnitude limit for short exposures in the low-resolution LR–K mode. The high-resolution mode has spectral channels which are 400 times narrower, and so an estimated correlated magnitude limit for short exposures in the HR–K mode is $K_{corr} \approx 6 - 2.5 \log_{10}(400) = -0.5$. Thus, the Mira observation would be marginal in this mode except on baselines where the source visibility is larger than the estimated value of 0.1.

Finally, the tip–tilt correction system on the VLTI has a magnitude limit of $V = 11$. Mira is highly variable in the V-band with a magnitude that oscillates between $V = 2$ and $V = 9.5$; even at its faintest the tip–tilt correction will work.

7.3 Wavelength and spectral resolution

The question of which waveband to observe a given object in depends on both the target physics and the characteristics of the available interferometric instruments. The aspects of the target physics which are important depend in part on the astrophysical question to be answered. For example, if velocity information or information about the emission from particular molecular or atomic species is required, the wavelengths to observe centre around relevant spectral emission or absorption lines in the source in question.

In other cases, the temperature of the feature of interest is relevant; for example, if it is required to make an image of the thermal emission from dust around a much hotter star, the contrast between the dust emission and the emission from the star will be most favourable at wavelengths near to the peak of the black-body curve at the temperature of the dust. For example, in dust discs around young stars, the inner rim of the dust will be at the sublimation temperature of the dust, perhaps 1500–2000 K. The peak of the black body in this case will be at a wavelength of around 2 μm, signifying that the near-infrared H and K bands are optimal. The cooler outer regions are better observed at longer wavelengths. Conversely, when scattering rather than thermal emission is to be observed, shorter wavelengths are often to be preferred.

Instrumental characteristics include the availability of suitable instruments working at a given wavelength; related to this is the sensitivity of these instruments. Few instruments work at the blue end of the visible spectrum because the sensitivity is so poor at these wavelengths (mainly due to the severity of the seeing) that there are few targets available. Similarly, there are few instruments targeting the mid-infrared because of the problems due to high levels of thermal background radiation.

Another instrumental factor is the availability of suitable baselines. If the highest angular resolution is needed this tends to favour using shorter wavelengths as the resolution is inversely proportional to the wavelength. Conversely, if the object is relatively large for the baselines available, going to longer wavelengths can help.

7.3.1 Mira example

In the case of the example observing proposal, the shocks are most evident in the emission lines of hydrogen. These lines have velocity dispersions of about 60 km s^{-1}, and so a spectral resolving power of 5000 would allow the line emission and the continuum to be separated. The hydrogen lines at visible wavelengths, for example Hα and Hβ, show a strong contrast with the stellar emission, but the only interferometric instrument with the required spectral resolution at visible wavlengths at the time was the VEGA instrument on CHARA, and the CHARA shortest baseline was around 30 m – much too long for objects of the size of Mira. Aperture masking on an 8-m class telescope using narrowband Hα filters would have been more appropriate but no system operating at the correct wavelengths was accessible for this programme. In the end the observing programme was targeted on the near-infrared Brγ line: although the contrast of this line with the continuum is poorer than for Hα, the HR–K mode of the AMBER instrument gave access to this line at an appropriate spectral resolution (R=12 000) and appropriate angular resolution for this target.

7.4 Baseline selection

Section 7.2 has already made the point that targets need to be selected based on the baseline range available; however, having selected the target, the choice of baselines within the range needs to be made. Ideally, one would observe the target on as many baselines as are available, but this can be prohibitive in terms of observing time and so the baselines which offer the most useful information should be prioritised.

The long baselines offer the highest resolution but the fringe visibility modulus tends to decrease with increasing baseline, and so if the baseline used is too long the SNR of the data may be too low to be of any use. Shorter baselines are needed to sample the low-spatial-frequency power in the object; a rule of thumb to estimate the shortest baselines to be included is given in Equation (7.2).

Adequate sampling of intermediate baselines is also required, depending on the field-of-view as given by Equation (7.3). With most current interferometers, achieving this level of sampling is difficult, but may not be necessary. As discussed in Chapter 9, it may be sufficient to sample the visibility at about N different points in the (u, v) plane where N is the number of free parameters in a physically plausible model of the source. In the case where the object can be modelled as a small number of point sources, N corresponds to the number of 'filled pixels' in the image; that is, the number of elements in an image pixellated at the resolution of the image which contain significant flux.

In such a case it is still essential to ensure that the N (u, v) samples are chosen to give a good diversity of coverage of the (u, v) plane in terms of both baseline length and the position angle of the baseline with respect to the source. Clearly, the use of Earth-rotation synthesis, i. e. observing the same source with a given pair of telescopes at different times of night, can help with this; choosing observation times spread evenly throughout the night can maximise the diversity with the minimum use of observing time: observations of multiple objects can be interleaved while waiting for the Earth to rotate.

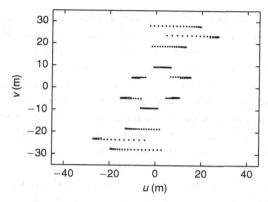

Figure 7.1 Baseline coverage tracks for observations of Mira (o Ceti) using the VLTI A0-B1-C2-D1 quadruplet of telescopes. The tracks shown are the coverage if Mira is observed at all elevations above 30°.

7.4.1 Mira example

The choice of baselines for the Mira observation was restricted by the available telescope positions for the semester of observations. A fixed number of layouts of the four VLTI auxiliary telescopes, known as 'quadruplets', were offered. Only the most compact of these, the A0-B1-C2-D1 quadruplet, allowed access to a baseline short enough to be inside the central lobe of the visibility function. It can be seen from Figure 7.1 that observations with this quadruplet will sample a good range of baseline lengths, but the position angle coverage of the (u, v) plane will be restricted. This means that circularly symmetric models can be well constrained, but making a model-independent image could be difficult.

7.5 Calibrator selection

A critical component of any interferometric observation is the observation of one or more calibrator stars immediately before and/or after the target observation. These stars serve to calibrate any systematic effects on the interferometric observables, particularly the power spectrum but also in some cases the closure phase and differential phase.

Calibrators need to have an object visibility function which can be predicted from existing data. Typically single stars with known angular diameters and limb darkening at the wavelength of observation are used. Stars with smaller angular diameters have the advantage that the visibility on a given baseline is a less strong function of the diameter, so any error in knowledge of the diameter or the limb darkening translates to a smaller error in the visibility.

Lists of interferometric calibrator stars for different wavelengths are now available (see Mérand *et al.*, 2005, and references therein) and these should be used where possible, as using a star which is not on these lists can be fraught with danger. A randomly chosen star has a high chance of being part of a multiple-star system and, depending on the brightness difference and angular separation, could have highly variable visibility on the observed baselines.

In addition to having a known visibility, an ideal calibrator should be as similar as possible to the science target in the parameters which are likely to affect the visibility. In order of priority these are:

1. Calibrators should be chosen to be close on the sky to the science target. This is beneficial for a number of reasons. The largest and most variable contributor to visibility degratation is the atmospheric seeing, so observing under as similar seeing conditions as possible is critical to good calibration. Seeing can vary with pointing direction, and varies systematically with distance from the zenith. Having nearby calibrators is particularly critical when the target is at low elevation, where the seeing degrades more rapidly with zenith distance. Mechanical vibrations such as wind-induced telescope vibrations can affect the visibility in the same way the seeing does and will also tend to be similar for pointing directions which are close to one another. The seeing and vibrations can also vary with time, and it is quicker to switch between the observation of a target and a nearby calibrator than one which is further away.

   An additional benefit of using close-by calibrators is that the delay-line positions and speed will be similar for the observations of the target and calibrator, and so any visibility-degrading phenomena (for example vibrations or diffraction), which depend on these factors, will be similar.

2. Calibrators should have similar brightnesses to the science targets in the science waveband. Non-linearities in the fringe detection process can lead to artefacts which depend on the total flux level. Also, many fringe analysis algorithms are non-linear and so the averaged fringe parameters could be degraded differently at different light levels unless care is taken to remove such systematic effects.

3. The wavelength dependence of the brightness of the calibrator should be similar to that of the science target, i. e. the colours should be similar. This aids in simultaneously matching the brightnesses at the science wavelength and at the wavebands at which the fringe tracker and AO systems work at. The performance of the active systems like the fringe tracker and AO system are light-level-dependent and so the level of residual phase perturbations will depend on the object brightnesses at the relevant wavelengths.

In addition, if a wide bandpass is being used for the science fringes, the effective wavelength of the fringes will be dependent on the colour of the object being observed and the visibility can be a function of this.

In reality, it is difficult to satisfy all these criteria at the same time and the choice of calibrators is based on a compromise between criteria, which requires judgement as to which are likely to be the more important sources of error. Using more than one calibrator for any given source confers a number of advantages. First, different calibrators can reflect different compromises between criteria. For example, choosing a close-by calibrator, which is different in brightness, and a further-away calibrator, which is more similar in brightness, allows the decision as to whether closeness or similarity in brightness is the more critical to be made after the fact, based on the actual data. Secondly, additional calibrators provide insurance against unexpected problems with a calibrator: for example, if a calibrator turns out to be binary, this can be checked against another calibrator and discarded if need be.

7.5.1 Mira example

Selection of the calibration stars for Mira was problematic because it is so bright. Stars which are of similar brightness are rare and therefore a long distance on the sky from Mira. Additionally, stars which are bright in the infrared are more likely to be cool stars: surface brightness constraints imply that such stars will have comparable angular diameters to Mira, and so the calibrator visibilities on the sampled baselines will be sensitive to any errors in the assumed calibrator diameter. The calibrators chosen were α Ceti and γ Eridani: both are fainter than Mira. The former is closer to Mira in brightness but further away in terms of angular separation while the latter is fainter but less distant and less resolved.

7.6 Surveys

Astronomy makes a lot of use of statistical analysis of multiple exemplars of a given class in order to infer information about the class or about relatively short-lived phenomena in the class. This technique can be equally powerful in interferometry, but does present problems in terms of the time taken to gather sufficient data. With most existing interferometers, gathering visibility information at a small number of (u, v) points on a single object can take anything from a few minutes to an hour, and sampling enough of the (u, v) plane to make a detailed image can take a large fraction of a night. With this kind of performance imaging surveys of more than a dozen objects or so can become impractical.

An alternative is not to try to image all the objects in the survey but rather to extract simple but scientifically important parameters, which require significantly less (u, v) coverage. One example is looking for binary stars, where in principle a single 'snapshot' with a small number of baselines may be sufficient to determine with reasonable confidence whether a system is binary (Le Bouquin and Absil, 2012). In such a case it may be possible to survey tens or even hundreds of stars in a reasonable time.

7.7 Short-timescale phenomena

Interferometry gives access to smaller physical scales than is possible with conventional imaging. This makes it likely that the phenomena that are observed on these scales are changing on shorter timescales, as the time taken for information to cross the object (at the speed of sound if the object is dense enough or at the speed of light if the object is transparent enough) or the time taken for orbital motion is correspondingly smaller.

To illustrate this principle one can consider a binary star system 10 parsecs away from the Earth: if it is 'visual binary' with a separation of order 1 arcsecond it will have an orbital period of around 30 years, while if it is an 'interferometric binary' with a separation of 10 milliarcseconds it will have a period of less than 9 hours. Another example is a nova occuring at a distance of 1 kiloparsec. If the ejecta from the star are travelling outwards at a speed of $1000 \, \text{km s}^{-1}$ then the diameter of the envelope will change by about 1 milliarcsecond per day.

Since interferometric observations can take significant fractions of a night or many nights if multiple telescope configurations are needed, it is possible that the object will change during the course of an observation. This can be taken into account if a model-fitting procedure with an explicit time-dependence is used, but making model-independent image reconstructions becomes much more problematic if object evolution is added as an additional dimension to the problem. Thus, evolution timescales must be taken into account when determining what kind of observation is feasible.

7.7.1 Mira example

The timescale of shock propagation in Mira was the main time constraint: if the shock is propagating radially from the star at speeds of $60 \, \text{km s}^{-1}$, the diameter of the shock front would change by about 5 milliarcseconds per week if Mira is at a distance of 100 parsecs, so observations within a few-day window could be considered as 'simultaneous'; observations separated by a few weeks would allow shock motion to be easily detected.

7.8 Complementary observations

The ultimate aim of any astronomical observation is to understand the physics of some particular phenomenon. The high-angular-resolution information provided by interferometry is typically critical in bringing new insights into the phenomenon that is the subject of an interferometric observing proposal, but the interferometric data can best be interpreted in the light of other information about the phenomenon. This information can often be gleaned by using lower-angular-resolution techniques such as spectroscopy or photometry, and by using information at different wavelengths, e. g. information about radio or X-ray emission.

While it is likely that some such general information already exists for the objects in the same class as those being observed, it is often the case that additional observations, such as photometry or spectroscopy of the targets in question, can provide valuable information. One example of this is the observation of binary stars, where spectroscopic information about the target system can in many cases be combined with interferometric measurements to derive all the physical parameters of the system.

Thus, the scientific value of an interferometric observation can often be maximised by considering it as just one observation in an observational campaign centred on a common scientific question, and making sure that observations on other telescopes are planned to provide any complementary information which is required to interpret the interferometric data.

Advance planning is especially important for time-variable objects where details about the object at the time of the interferometric observation can be critical. For example, Mira variables show cycle-to-cycle variations in many of their properties and so observations which are within the same cycle as the interferometric observations may prove valuable.

8

Data reduction

The 'raw' data from an interferometer consist of the measurements of the fringe pattern plus auxilliary data required for calibration. These data need to be converted into calibrated power spectrum and bispectrum data or coherently-average visibility data for subsequenty model-fitting and image reconstruction. The exact details of the data-reduction process varies between interferometric instruments and typically software is provided for each instrument that can perform the major parts of the process. This chapter provides an outline of what is going on inside this software in order to provide an understanding of the processes and the rationale for choosing one process over another when analysing a given dataset.

8.1 Scientific inference

The data-reduction process is part of a larger process, which aims to gain some knowledge about the astronomical object under investigation based on measurement of fringe patterns, and it is helpful to consider the process as a whole to understand where data reduction fits in.

The process of gaining knowledge based on measurements is known as *scientific inference*. A conceptual model of scientific inference starts out with an existing state of knowledge. This can be cast in terms of a model of the object, which has a number of unknown parameters. An example model is a binary star system consisting of a pair of stars with unknown brightnesses and diameters for the constituent stars and an unknown separation between them.

A particular set of values for all the model parameters can be thought of as representing a single point in a multi-dimensional space known as the *model space*. For a particular point in model space, the set of fringe measurements that would be produced by a given interferometer represents a point in the *data*

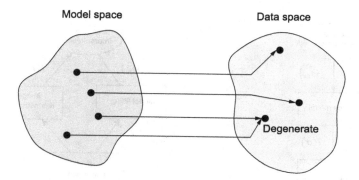

Figure 8.1 Schematic outline of the forward problem in scientific inference.

space of the instrument. This model of inference is shown diagramatically in Figure 8.1.

Computing the point in data space corresponding to a given point in model space is relatively straightforward and is known as the *forward problem*. The reverse process, that of taking a point in data space representing a set of measurements and inferring where in model space the data came from, is known as the *inverse problem* and is generally more difficult.

One difficulty is brought about by the presence of noise, which means that a single set of model parameters can lead to multiple possible measurement values. The problem may also be *degenerate* as shown in Figure 8.1: even in the absence of noise, multiple different sets of model parameters could result in the same values for the measured data. These difficulties mean that there may be no unique solution to the inverse problem, but instead a 'space' of possible solutions must be considered.

Solving the inverse problem lies at the heart of scientific inference, since the aim of the scientific method is to discern something about an unobservable model of the world based on its observable effects.

8.2 The forward problem

Understanding how to solve the inverse problem for an interferometer starts with studying the forward problem and this needs a model which includes both the object and the measurement process. The model used here attempts to balance generality against simplicity and so does not include all the important features of all interferometeric instruments. Nevertheless, it should provide a useful starting point for understanding the data-reduction process.

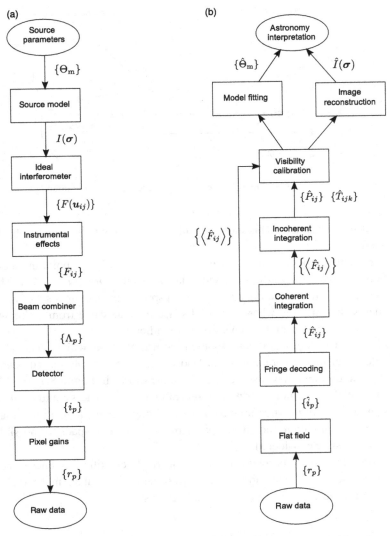

Figure 8.2 The forward model of an interferometric measurement (a) and an outline of the data reduction process for interferometric data (b).

A flow diagram of the forward model is shown in Figure 8.2(a) and starts with a model for the object under study. The details of this model are clearly dependent on the particular object and so it is represented as an abstract model with a set of parameters $\{\Theta_m\}$.

Assigning a particular set of values to the object parameters corresponds to selecting a single point in the multi-dimensional model space. From the object parameter values the model will predict a brightness distribution on the sky $I(\sigma)$. This brightness distribution is observed by an idealised interferometer, which samples the object coherent flux at known locations $\{u_{ij}\}$ in the (u, v) plane, yielding coherent fluxes $F(u_{ij})$.

Instrumental and atmospheric effects will disturb the visibility modulus and phase from their ideal values. These disturbances are represented by introducing complex fringe degradation factors γ_{ij}, which relate the fringes seen in an idealised interferometer to those observed in the presence of these effects. The coherent fluxes of the measured fringes are given by

$$F_{ij} = \gamma_{ij}F(u_{ij}). \tag{8.1}$$

The beam combiner converts the coherent fluxes F_{ij} into a fringe pattern, which is represented as a set of discrete intensity samples $\{\Lambda_p\}$ where Λ_p is the integrated classical intensity falling on the detector as defined in Section 5.2. These samples could be pixel values at different spatial locations on a detector or temporal samples when fringes are scanned across a single pixel, or a mixture of the two. The individual samples will be called 'pixels' whether or not the samples actually come from a two-dimensional detector.

The samples are divided into a series of independent *interferograms* each consisting of N_{pix} samples. For temporally coded fringes, the interferogram could consist of a single 'sweep' of the pathlength modulation or it could consist of a subset of the sweep which contains sufficient data to extract the fringe parameters.

The sampled intensities $\{\Lambda_p\}$ are assumed to be related to the coherent fluxes F_{ij} by a linear superposition of 'carrier waves' (or 'coding waveforms') w_{ijp} so that

$$\Lambda_p = \sum_{i \leq j} \text{Re}\left\{F_{ij}w_{ijp}\right\}, \tag{8.2}$$

The quantity F_{jj} represents the flux from an individual telescope and so w_{jjp} corresponds to the shape of the 'background' intensity pattern while w_{ijp} for $i \neq j$ corresponds to the modulation of the fringes on baseline ij. Note that the carrier waveform implicitly includes any pixel-to-pixel variations in the quantum efficiency of the detector, so that the intensity Λ_p is in units of photo-events.

By choosing appropriate forms for w_{ijp} most beam combiners can be modelled using the *fringe composition equation* given in Equation (8.2).

For example, to model an ideal two-telescope interference pattern consisting of a uniformly sampled sinusoidal one-dimensional fringe pattern with a uniform background, the waveforms can be represented by $w_{11} = w_{22} = \frac{1}{2}$ and $w_{12p} = e^{2\pi i s p}$, where s is the spatial frequency of the fringes expressed in cycles per pixel so that

$$\Lambda_p = \tfrac{1}{2}(F_{11} + F_{22}) + \mathrm{Re}\left\{F_{12} e^{2\pi i s p}\right\}. \tag{8.3}$$

The detector converts the intensity values into electronic levels, which are digitised. The detection process is modelled as two independent steps, whereas in reality these steps are coupled to one another. First, the noise-free interferogram Λ_p is corrupted by zero-mean additive noise n_p, yielding noisy values i_p given by

$$i_p = \Lambda_p + n_p. \tag{8.4}$$

In the second step these values are multiplied by a pixel-dependent gain g_p (known as a 'flat field') and a systematic background offset b_p is added to give the 'raw' data values:

$$r_p = g_p i_p + b_p. \tag{8.5}$$

Non-linearity of the detector response could also be included in the model but is ignored here for simplicity.

8.3 The inverse problem

The inverse problem for interferometry is to take a set of raw intensity measurements $\{r_p\}$ and use these to learn something about the model of the object that produced these measurements in terms of the values of the parameters $\{\Theta_m\}$. One method for solving inverse problems is known as Bayesian inference, which is explained in more detail in Section 9.1.

A fully Bayesian solution to the inverse problem involves fitting the complete forward model of the system, which includes the object unknowns and all the instrumental unknowns such as the atmospheric wavefront effects and noise to the all of the raw data and any additional calibration data, to compute a multi-dimensional probability distribution of the model parameters. The inference process described here is not fully Bayesian; it is composed from a number of separate steps most of which are non-Bayesian. A similar multi-step process is used for analysing data at most interferometers, because it is mathematically and computationally more tractable for dealing with large amounts of data than a fully Bayesian solution.

Many of the steps in the data reduction process rely on the use of statistical *estimators*. An estimator is a function which maps a single point in data space (i.e. the actual set of measurements) to a single point in model space. This point represents an estimate for one or more model parameters based on the data. This can be contrasted with the Bayesian method, which uses a single set of measurements to derive probability estimates for every possible set of model parameter choices. The choice of which estimator to use for a given problem is somewhat ad-hoc, but well-chosen estimators can give results similar to those given by Bayesian analyses.

A schematic diagram of the inference process described here is shown in Figure 8.2(b), which starts from the raw data and computes a number of intermediate quantities before ending up with either an image of the object or a set of model parameters for the object. These steps are discussed briefly below and then expanded upon in later sections.

The first step of the process is to remove detector artefacts such as non-uniform flat fields and background levels to produces a set of estimates of the noisy intensities $\{\hat{i}_p\}$ (a circumflex is used distinguish the value of the estimator from the 'true' value of i_p). The intensity estimates are then used to derive estimates of the coherent flux $\hat{F}_{ij}$ on an interferogram-by-interferogram basis.

The coherent fluxes can be coherently integrated over a number of exposures to increase the signal-to-noise ratio (SNR) of the estimate. The coherent-flux estimates then are converted into power spectrum and bispectrum estimates, $\{\hat{P}_{ij}\}$ and $\{\hat{T}_{ijk}\}$, which are independent of the effects of atmospheric piston errors.

The data can be 'edited' at this stage to remove outliers before being incoherently averaged. The incoherently averaged power spectrum and bispectrum data are then calibrated for the effects of residual visibility corruption factors using data from observations of calibrator stars.

The final step is to use the calibrated data to constrain models or to reconstruct images of the object. This step is the subject of the next chapter; the data reduction process described in this chapter describes all the stages needed to go from the raw intensity data to the calibrated and incoherently averaged power spectrum and bispectrum data.

8.4 Flat fielding and background subtraction

The process for compensating for the detector gains g_p and offsets b_p relies on having accurate measurements for these values. These calibrations are usually in the form of 'darks' (i.e. exposures with the detector shuttered) and 'flats' (exposures where the detector is uniformly illuminated). The darks and flats can be analysed to produce estimates for $\hat{b}_p$ and $\hat{g}_p$, background and gain, respectively. The exact way these calibration measurements are taken and used are instrument-dependent and so further details for this process are not given here. Given these estimates, a set of corrected intensity values can be computed as

$$\hat{i}_p = \frac{\left(r_p - \hat{b}_p\right)}{\hat{g}_p}. \tag{8.6}$$

Note that, as defined in Equation (8.4), the units of i_p are the same as the units for Λ_p, i.e. detected photo-events, so the estimates for the gain need to be normalised on this basis. Scaling errors in this normalisation can affect the bias-correction stage (see Section 8.7).

8.5 Extracting the coherent flux

8.5.1 Sinusoidal model

In most beam combiners the carrier waveforms w_{ijp} for the fringes are sinusoids with slowly varying envelopes. For a one-dimensional pattern where the fringes are sampled in equal increments of phase, the waveforms can be represented as

$$w_{ijp} = a_{ijp}e^{2\pi i s_{ij}p} \tag{8.7}$$

where s_{ij} is the frequency of the fringes caused by interference between telescopes i and j and a_{ijp} is the envelope function for those fringes. Typically, $s_{jj} = 0$ so that $w_{jjp} = a_{jjp}$, representing just a slowly varying background offset to the fringe pattern caused by the non-interfering light.

An example fringe pattern is shown in Figure 8.3. This fringe pattern consists of the superposition of sinusoids at three different spatial frequencies, each with a Gaussian envelope of a different width, on a flat background (i.e. w_{jjp} is constant for all p). This could represent, for example, the data from a temporally scanned beam combiner where the envelopes are due to longitudinal coherence effects (see Section 1.7) reducing the contrast of the fringes.

8.5.2 The discrete Fourier transform

With this data model it is natural to consider using a Fourier transform to extract the sinusoidal components from the data. The appropriate transform to use with discrete data is the discrete Fourier transform (DFT).

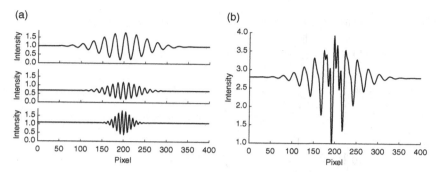

Figure 8.3 A discretely sampled fringe pattern with an envelope: individual fringe patterns (a) and the superposed pattern (b).

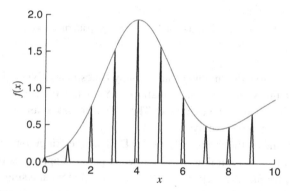

Figure 8.4 Discretising a continuous function represented as converting it to a series of Dirac delta functions.

For one-dimensional data the DFT is given by

$$I_k = \sum_{i=0}^{N_{\text{pix}}-1} i_p e^{-2\pi i k p/N_{\text{pix}}}, \tag{8.8}$$

for $k = 0, 1, \ldots, N_{\text{pix}} - 1$. Equivalent two-dimensional discrete transforms are a natural extension of this but are not discussed in detail here.

The DFT can be understood as a conventional continuous Fourier transform in which the function to be transformed is represented as set of delta functions with unit spacing whose magnitudes (i. e. areas) are given by the sample values as shown in Figure 8.4. The DFT is then the continuous Fourier transform sampled in frequency space at intervals of $1/N_{\text{pix}}$ cycles per pixel.

The DFT is most commonly implemented on the computer using an algorithm known as the fast fourier transform (FFT), which, as its name suggests,

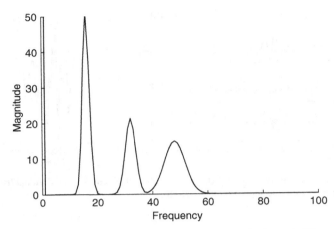

Figure 8.5 The magnitude of the DFT of the fringe pattern shown in Figure 8.3.
The component at zero frequency has been truncated.

impements the DFT in a manner that is many times faster than a direct implementation of the summation in Equation (8.8) if the Fourier coefficients at
many frequencies are to be extracted. The FFT also has a numerical accuracy
which is superior to more naïve methods.

Figure 8.5 shows the magnitude of the DFT of the fringe pattern shown in
Figure 8.3. It can be seen that the three fringe patterns appear as peaks at different spatial frequencies. These peaks do not just appear at a single frequency
but are broadened, with the broadening being different for different peaks.

The reason for this broadening can be understood from the convolution theorem. The component waveforms are the product of an envelope a_{ijp} and a
sinusoidal component $e^{2\pi i s_{ij} p}$ and so the Fourier transforms of each waveform
will be the convolutions of their Fourier transforms. The Fourier transform of
the sinusoid will be a delta function at frequency s_{ij} and the transform A_{ijk}
of the Gaussian envelope a_{ijp} will be a Gaussian whose width in the DFT is
inversely proportional to the width of the Gaussian envelopes in the fringe
patterns.

The value of the DFT at the discrete frequency k/N_{pix} which is closest to s_{ij}
can be taken as an estimator for F_{ij}. More sophisticated approaches are possible, for example interpolation of the DFT values, but these are not considered
here. If s_{ij} is not known a priori, then it can be estimated from the DFT by
looking for the frequency at which the Fourier magnitude is maximised for
each peak in the spectrum.

The zero-spacing fluxes F_{ii} all contribute to the peak at the origin of the
DFT, so if the envelopes a_{ii} are similar in shape then the individual fluxes

cannot be computed from the fringe pattern, only their sum. If it is desired to compute the individual fluxes, additional photometric channels are needed.

8.5.3 The *ABCD* estimator

The DFT estimator is widely used in interferometry. One particularly simple form of the DFT is the so-called *ABCD* estimator, This has been used in a number of interferometers which scan the fringes, such as the Mark III interferometer on Mt Wilson (Shao *et al.*, 1988) and the PTI (Colavita *et al.*, 1999).

This estimator is used when the fringe is sampled at intervals of $\pi/2$ in phase, and four successive samples are used to determine the fringe amplitude and phase. Labelling the four intensities as $i_0 = A$, $i_1 = B$, $i_2 = C$, $i_3 = D$, the fringe appears in the DFT at a frequency of $1/4$ so that

$$\hat{F}_{ij} = \sum_{p=0}^{3} i_p e^{-ip\pi/2} \tag{8.9}$$

$$= A - C + i(B - D). \tag{8.10}$$

Thus, the real and imaginary parts of the coherent flux are easily computed, and the amplitude and phase of the fringes can be computed as

$$|F_{ij}| = \sqrt{(A - C)^2 + (B - D)^2} \tag{8.11}$$

and

$$\arg F_{ij} = \arctan\left(\frac{B - D}{A - C}\right), \tag{8.12}$$

respectively.

8.5.4 Spectral leakage and windowing

Fringe-parameter estimators based on the DFT work well when the data can be accurately modelled by Equation (8.7). However, there are a number of practical situations where this model is not an accurate one.

One such situation is where the data are not evenly sampled in phase. This could be because the fringes are temporally scanned and the scan velocity is not constant, as might occur for a mechanical sawtooth-like scan when the scanning mirror is 'turning around' at the ends of the scan. Another cause for this uneven sampling could be if there are 'dead' pixels on a detector, which means that some of the samples are missing.

Both of these situations can be modelled by allowing the fringe envelope a_{ijp} to be a complex number, whose phase adjusts the phase of the sinusoid in an appropriate way to account for the uneven phase sampling. However, this will likely run into a second situation in which the DFT does not work well, which is when the fringe envelope is not sufficiently smooth and so 'spectral leakage' occurs.

As discussed in Section 8.5.2, the Fourier transform of the fringe pattern is the sum of a set of delta functions at the frequencies of the fringes s_{ij}, convolved with the Fourier transform A_{ijk} of the fringe envelopes. The Fourier transform of the fringe envelope can therefore be thought of as 'scattering' power from the fringe frequency to the neighbouring frequencies. If the envelopes are sufficiently smooth, A_{ijk} will be compact, and the fringe peaks will be well separated in the DFT. If the envelopes have high-frequency structure so that A_{ijk} is broad or if neighbouring fringe frequencies are too close together, then it is possible for the peaks to overlap and power to be scattered from one fringe peak into another, or from the peak at zero frequency to the lowest fringe frequencies.

A common example of such high-frequency structure is the sharp truncation of a sinusoidal fringe pattern due to sampling the pattern over a finite interval. If a sinusoidal fringe pattern is sampled over a integral number of fringe cycles, the DFT of this pattern will have a single peak at the fringe frequency. If there are a non-integral number of cycles, then the peak will be broadened so there is significant power at a range of frequencies as shown in Figure 8.6.

This effect can be seen as being due to the multiplication of an infinitely long sinusoid by a 'top-hat' function representing the finite length of the sampling. The Fourier transform of a top-hat of length T is a sinc function with zeros at frequency intervals of $1/T$. If the fringe frequency s_{ij} is an integer multiple of $1/T$, then the convolution of the sinc function with a delta function at s_{ij} has a zero at the origin, meaning there is no 'cross-talk' between the fringe and the zero-frequency level.

If the fringe frequency is not an integer multiple of $1/T$ then there is cross-talk arising from the sidelobes of the sinc function as shown in Figure 8.6. A similar cross-talk effect will also occur between neighbouring fringe frequencies if they are not separated in frequency by integral multiples of $1/T$.

This spectral leakage of power between nearby fringe frequencies due to truncation of the fringe pattern can be reduced by a technique known as 'windowing' or 'tapering'. This technique involves multiplying the intensity pattern $\hat{\imath}_p$ by a 'window function', which is tapered at the edges.

One example of a window function is a Gaussian whose values at the edges of the data are close to zero as shown in Figure 8.7. Multiplying by the window

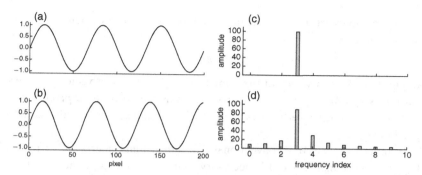

Figure 8.6 A sinusoid with an integral number of fringe cyles over the sampling interval (a) and with a non-integral number of fringe cycles (b) and their corresponding discrete Fourier transforms (c, d). The DC value of the fringe pattern has been set to zero to provide greater clarity.

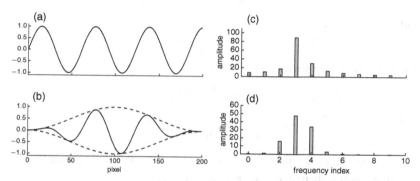

Figure 8.7 Windowing by a Gaussian (a, b) and its effect on spectral leakage (c, d).

function smooths the discontinuity at the edge of the sinusoid caused by truncation; in Fourier space, it corresponds to convolving the DFT with a function whose 'wings' are much more rapidly decaying than the sidelobes of the sinc function, and so the leakage to nearby frequencies is reduced.

8.5.5 The pixel-to-visibility matrix

Windowing only partially solves the problem of spectral leakage, and does so at the expense of SNR as it 'weights down' the data at the edges. In addition, there are other effects, such as uneven sampling, which cause spectral leakage and which cannot be solved by windowing. An alternative estimator that completely avoids spectral leakage can be found using matrix methods.

This method starts from noting that Equation (8.2) can be written in the form of a matrix equation,

$$\Lambda = Mf, \tag{8.13}$$

where $\Lambda = \Lambda_p$ is a vector of pixel intensities, $f = \{f_b\}$ is a vector of coherent fluxes and $M = \{M_{pb}\}$ is a matrix known as the *visibility-to-pixel matrix* or *V2PM*, since it converts visibilities (strictly speaking, coherent fluxes) into pixel intensities.

In order to make M a conventional two-dimensional matrix rather than a higher-dimensional tensor it is necessary to map the telescope index pairs to a unique one-dimensional baseline index $\{i, j : i, j = 1, \ldots, N_{tel}\} \rightarrow \{b : b = 1, \ldots, N_{tel}^2\}$. In order to make M and f real, it is necessary to also map the telescope pairs to a second index b' so that

$$f_b = \text{Re}\left\{F_{ij}\right\} \tag{8.14}$$

and

$$f_{b'} = \text{Im}\left\{F_{ij}\right\}. \tag{8.15}$$

The V2PM can then be written

$$M_{pb} = \text{Re}\left\{w_{ijp}\right\} \tag{8.16}$$

and

$$M_{pb'} = -\text{Im}\left\{w_{ijp}\right\}. \tag{8.17}$$

Given this representation of the forward equation as a matrix equation, the question arises if there exists an inverse matrix H such that

$$\hat{f} = H\Lambda \tag{8.18}$$

provides a 'good' estimate for the coherent fluxes f.

Since the V2PM is in general not square (but should rather be overdetermined, i. e. there should be more pixels than visibilities) there exists no unique inverse matrix. However, there does exist a *pseudoinverse* often called the Moore–Penrose pseudoinverse. This matrix has the property that it gives a least-squares solution to the problem; in other words, given a vector of noisy intensities $i = \{\hat{i}_p\}$, it provides a vector of coherent fluxes,

$$\hat{f} = Hi, \tag{8.19}$$

which minimises the squared difference χ^2 between the measured intensities and the intensities that would be predicted from the estimated coherent fluxes. In other words,

$$\chi^2 = |i - MHi|^2 \tag{8.20}$$

is minimised. This least-squares estimate is an optimal estimate of f from intensity measurements dominated by Gaussian read noise which is the same on all pixels.

The Moore–Penrose pseudoinverse can be computed from the V2PM by singular value decomposition (SVD), which comes as a standard package in many numerical libraries. The pseudoinverse in this case is is called the *pixel-to-visibility matrix* or *P2VM* since it converts pixel intensities to coherent fluxes or 'visibilities'.

The P2VM H can be computed providing that the V2PM M is known. This technique is used in the AMBER instrument on the VLTI (Tatulli *et al.*, 2007). As part of the AMBER calibration sequence, light is injected into the single-mode fibres to form fringes with known F_{ij} and this is used to estimate the matrix M, and hence to estimate H. AMBER also has photometric channels, which allow the fluxes from individual telescopes F_{ii} also to be estimated using a P2VM. In other instruments, it may not be possible to estimate the individual fluxes but only their sum.

The P2VM estimator is less directly applicable if the matrix M is unknown. This can occur, for example, if the carrier waveforms w_{ijp} change shape due to the effects of atmospheric turbulence. Nevertheless, it may still be possible to determine a suitable estimator for f, and if this estimator is linear it can be represented by a matrix.

An example of this is that the DFT method is robust to changes of shape in the envelope a_{ijp}, provided the envelope never has components at sufficiently high spatial frequencies that fringe power is scattered into nearby fringe components. The DFT estimator can be represented as a matrix H consisting of sinusoidal components

$$H_{bp} \propto \cos(2\pi s_{ij} p) \qquad (8.21)$$

and

$$H_{bp} \propto -\sin(2\pi s_{ij} p). \qquad (8.22)$$

In the following, it will be assumed that a linear estimator is used to estimate the coherent fluxes and that this estimator is represented by a P2VM H, independent of whether or not the matrix was computed using a direct pseudoinverse of M.

The P2VM formalism is a very general one. There is no requirement that the pixels are contiguous, so it can be applied to beam-combining systems where the fringe patterns for different baselines are superposed on a single camera, where the fringe patterns for different baselines are imaged by different cameras or where the information for a single baseline is extracted from more than one camera, as may happen when complementary outputs of a beamsplitter-based combiner are used.

8.6 Coherent integration

8.6.1 Passive integration

The coherent fluxes estimated from a single interferogram may be noisy, and it was shown in Chapter 5 that at low light levels the power spectrum and bispectrum derived from a coherent flux with an SNR of less than one will be even more noisy. From this point of view it would be attractive to try to increase the SNR of the coherent flux estimate by averaging multiple exposures.

The simplest way of performing this *coherent integration* is to integrate the pixel data over N exposures.

$$\hat{i}'_{p,k_0} = \sum_{k=k_0}^{k_0+N} \hat{i}_{p,k}, \qquad (8.23)$$

where $\hat{i}_{p,k}$ represents the intensity estimate in pixel p on exposure k. This is essentially equivalent to taking longer exposures (except that there is more readout noise because there are more readouts), and so it suffers from the same problem as longer exposures – if the effective integration time is comparable to t_0 the fringes will smear out and therefore have lower visibility. This will lower the SNR of the fringe measurement and will tend to cancel out the intended effect of coherent integration.

8.6.2 Phase-referenced averaging

If there is a source of information as to how the fringes are moving between exposures, this can be used to compensate for the effects of fringe motion during a coherent integration using a technique known *phase-referenced averaging* or *software cophasing*.

The information about the fringe motion can come from several sources. One possibility is that the information comes from a dedicated fringe tracker. As described in Section 3.8, fringe trackers are typically used for *hardware cophasing*, in which the fringe sensor measures the atmospheric piston changes and compensates for these changes in real time using a movable mirror or similar hardware compensation element. This compensation will not be perfect, because the piston compensator will take a finite time to move to the required location, during which time the fringe will have moved. An alternative to hardware cophasing (or something which can be used in conjuction with hardware cophasing) is to use the fringe tracker measurements for post facto compensation of the science combiner fringe phases in software. This can offer an improvement over hardware cophasing because software cophasing suffers from no 'lag' between phase measurement and correction. Indeed, because the correction is done in post-processing, the correction can make use of knowledge of the phase motion in the future as well as the past of the exposure being corrected.

Additional fringe-tracking information can come from the science combiner itself. Coherent integration is of value when the SNR of the fringes in a given science channel is low, and under these circumstances the data from that science channel has too low an SNR to provide the information for fringe tracking. There may, however, be other channels in the science combiner which do have higher SNRs. One example is if a wide-band channel is used to provide fringe tracking for a narrow-band channel. In a spectrally dispersed fringe pattern the wideband channel can be synthesised by adding together the data from all the narrow-band channels.

A common occurrence is that the fringe visibilities on different baselines can be widely different, and this can lead to a situation where two high SNR baselines form a closing triangle with a third baseline with much lower SNR. In this case, the phase changes seen on the two high-SNR baselines can be used to infer the phase changes on the low-SNR baselines, since the closure phase is constant – this is a form of 'baseline bootstrapping'.

Given an estimate of the fringe motion over a number of exposures, the fringe data can be compensated for this motion using a technique known as *phase rotation* or *fringe rotation*. To do this, the coherent flux $\hat{F}_{ij,k}$ is estimated for each exposure and then the phase of this coherent flux is adjusted based on an estimate ϕ_k of how the phase of the fringes in that exposure has changed compared to some reference exposure. This involves multiplying the visibility by a complex phasor

$$\hat{F}'_{ij,k} = e^{-i\phi_k} \hat{F}_{ij,k}, \qquad (8.24)$$

which rotates the complex vector $F_{ij,k}$ in the Argand diagram.

Assuming that the phase change estimate ϕ_k is correct then the phase-rotated coherent fluxes for a sequence will all have the same phase. They can then be added together with no loss of visibility and a corresponding increase in SNR. If the read noise from the science combiner detector can be neglected, then the resultant sum has the same properties as the visibility from a long-exposure fringe measurement made with hardware cophasing.

Time-varying errors in the fringe-motion estimates can arise for a number of reasons, including low SNRs caused by periods of bad seeing ('drop-outs') and time-variable atmospheric chromatic dispersion effects between the fringe-tracking wavelength and the science wavelength. The longer the sequence of phase estimates, the greater the chance that there will be an error in the estimate of the fringe motion, which will cause a loss in fringe contrast, so coherent integration is often limited to the averaging of relatively short sequences of data. On the other hand, if it is possible to coherently integrate an entire sequence of data (Jorgensen *et al.*, 2010) then it may not be necessary to perform incoherent integration.

8.7 Incoherent integration

The coherently integrated data can be summarised by averaging quantities which are independent of any phase perturbations, such as the bispectrum and power spectrum.

8.7.1 Power spectrum

In Section 5.5.2 it was shown that a naive estimator for the power spectrum is biased in the presence of detection noise and that an unbiased estimator must be used. In deriving this estimator it was assumed that a DFT was being used to derive $\hat{F}_{ij}$ and that the noise was additive Gaussian noise, which was independent of the signal level in a given pixel. These assumptions can be relaxed to include the effects of Poisson noise (i. e. photon noise) and to allow a more general P2VM to be used at the expense of more complex mathematics (Gordon and Buscher, 2012).

The results from such an analysis are broadly in line with the results from the Gaussian model. The bias-free estimator based on a coherent flux estimator $\hat{F}_{ij}$ in the presence of photon noise and Gaussian read noise of σ_p on pixel p is

$$\langle \hat{P}_{ij} \rangle = |\hat{F}_{ij}|^2 - \sum_p \left(\hat{i}_p + \sigma_p^2 \right) |H_{bp}|^2 , \qquad (8.25)$$

where H_{bp} is the P2VM used in estimating $\hat{F}_{ij}$.

If there are uncertainties in the calibration of $\hat{i}_p$ in units of photo-events or in the value of the read noise σ_p then there may be difficulties in estimating the bias correction terms in Equation (8.25). In many cases a DFT is used for the P2VM and there are frequencies at which there is expected to be no fringe power. In these cases the power spectrum bias can be estimated from the 'background level' of the power spectrum at these freqeuncies, since for a DFT $|H_{bp}| = 1$.

8.7.2 Bispectrum

If the coherent flux F_{ij} is estimated using a DFT and the detection noise is purely Gaussian, there is no bias correction term on the bispectrum. However, Poisson noise has a non-zero third-order moment and so it can induce a bias on the bispectrum. In addition, using a non-DFT P2VM can also lead to biases. A bias-free estimator in the case of a general P2VM H and a mixture of photon noise and Gaussian read noise is (Gordon and Buscher, 2012).

$$
\hat{T}_{123} = \hat{F}_{12}\hat{F}_{23}\hat{F}_{31}
$$
$$
- \hat{F}_{12} \sum_p \left(\hat{i}_p + \sigma_p^2\right) H_{23,p} H_{31,p}
$$
$$
- \hat{F}_{23} \sum_p \left(\hat{i}_p + \sigma_p^2\right) H_{12,p} H_{31,p}
$$
$$
- \hat{F}_{31} \sum_p \left(\hat{i}_p + \sigma_p^2\right) H_{12,p} H_{23,p}
$$
$$
+ \sum_p \left(2\hat{i}_p + 3\sigma_p^2\right) H_{12,p} H_{23,p} H_{31,p}. \tag{8.26}
$$

When the DFT is used so that $H_{ij,p} = e^{-2\pi i s p}$ then a simpler form is recovered

$$
\hat{T}_{123} = \hat{F}_{12}\hat{F}_{23}\hat{F}_{31}
$$
$$
- \left|\hat{F}_{12}\right|^2 - \left|\hat{F}_{23}\right|^2 - \left|\hat{F}_{31}\right|^2 + 2N_{\text{phot}} + 3N_{\text{pix}}\sigma_{\text{pix}}^2, \tag{8.27}
$$

where it has been assumed that the noise on all N_{pix} pixels has the value σ_{pix} and N_{phot} is the total number of photons given by

$$
N_{\text{phot}} = \sum_{p=0}^{N_{pix}-1} \hat{i}_p. \tag{8.28}
$$

It is important for accurate bias subtraction in the bispectrum that the intensities $\{\hat{i}_p\}$ are calibrated in units of photo-events and that the read noise is well characterised.

8.8 Visibility calibration

8.8.1 Antenna-based gains

The forward model includes a visibility corruption factor γ_{ij} due to atmospheric and instrumental effects, which can cause large discrepancies between the object visibility function $F(u_{ij})$ and the measured fringe parameters F_{ij}.

Some of the degration effects on the fringes can be represented by associating a complex 'gain factor' η_i with each arm of the interferometer (in radio interferometry these factors are known as *antenna-based gains*), so that

$$\gamma_{ij} \approx \eta_i \eta_j. \tag{8.29}$$

For example, piston wavefront errors ϵ_i associated with the atmosphere above a telescope i and the beam train from the telescope to the beam combiner can be incorporated as an antenna-based gain $\eta_i = e^{i\epsilon_i}$. Light losses in single-mode fibre beam combiners can be represented as values of $|\eta_i|^2$ that are equal to the coupling efficiency of light power into the fibre.

The antenna-based phase errors cancel out in the closure phase, and so no further calibration is needed to remove the effects of these errors. In radio interferometry, quantities known as 'closure amplitudes', which involve multiplying and dividing the magnitudes and coherent fluxes measured on four or more baselines are sometimes used to cancel the errors caused when $|\eta_i| \neq 1$, but these have not gained a foothold in optical interferometry.

8.8.2 Photometric calibration

An alternative way of compensating for the errors in the measured visibility modulus due to antenna-based gain errors is to use so-called *photometric calibration* methods. These methods rely on measuring the fluxes from individual telescopes F_i and F_j by 'peeling off' a small fraction of the light after it has undergone coupling into a single-mode system but before beam combination, as shown in Figure 8.8. These 'photometric channels' allow the measurement of the instantaneous coupling efficiencies since $F_i = |\eta_i|^2 F(0)$ where $F(0)$ is the flux from the object, and the product of these efficiencies is related to the coherent flux. The simplest way to use these photometric channels is to estimate the root-mean-square (RMS) visibility using the ratio

$$\hat{V}_{\text{RMS},ij} = \sqrt{\frac{\langle \hat{P}_{ij} \rangle}{\langle \hat{F}_i \hat{F}_j \rangle}}$$

$$= \sqrt{\frac{\langle |\eta_i \eta_j| \rangle |F(\boldsymbol{u}_{ij})|^2}{\langle |\eta_i|^2 F(0)|\eta_j|^2 F(0) \rangle}}$$

$$= |V(\boldsymbol{u}_{ij})|. \tag{8.30}$$

Thus, the dependence of the visibility on the antenna-dependent terms has been eliminated. Single-mode beam combiners with photometric channels such as

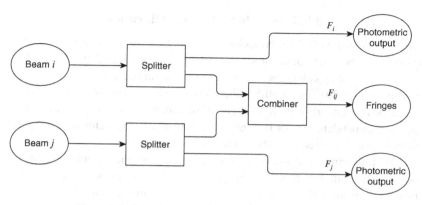

Figure 8.8 Schematic of a photometric calibration system.

the VINCI instrument on the VLTI have produced visibilities that have been calibrated at the 1% level or better (Ségransan *et al.*, 2003).

The fraction of the flux which is extracted in the photometric channels is usually quite small (a few percent in many cases) and so the instantaneous SNR of $\hat{F}_i$ may be less than unity for faint sources. In such a case the product term in the denominator $\hat{F}_i\hat{F}_j$ may have a low SNR, even after averaging over many frames. In these circumstances a preferable estimator to use might be

$$\hat{V}_{\text{RMS},ij} = \sqrt{\frac{\langle P_{ij}\rangle}{\langle \hat{F}_i\rangle\langle \hat{F}_j\rangle}}. \tag{8.31}$$

Using the methods employed in Section 3.7, it can be shown that

$$\langle \hat{F}_i\rangle\langle \hat{F}_j\rangle \approx \langle \hat{F}_i\hat{F}_j\rangle, \tag{8.32}$$

providing that the fluctuations in the coupling coefficients $|\eta_i|^2$ and $|\eta_j|^2$ are uncorrelated. This assumption will be true over short periods where the coupling fluctuations are due to changes in the instantaneous wavefronts across each telescope, but the fluctuations can become correlated if there is a systematic change in r_0 across both telescopes during the averaging period. Thus, it is best to apply the estimator given in Equation (8.31) to averages over timescales that are short compared to the expected timescale of any fluctuations in the value of r_0.

8.8.3 Transfer-function calibration

Many sources of visibility corruption, for example those due to fringe smearing during an exposure or polarisation mismatches, are baseline-dependent factors, that is to say they cannot be factorised into antenna-based gains as assumed in Equation (8.29). These effects therefore cannot be removed using closure-phase or photometric calibration. The most common way of calibrating these baseline-dependent factors is to use observations of 'calibrator stars' and this technique is described in outline in this section.

The following discussion will concentrate on the calibration of the power spectrum but most of the principles can be carried over straight-forwardly from calibration of the power spectrum to calibration of the bispectrum.

Calibration of the power spectrum at visible and near-infrared wavelengths is usually carried out in terms of the visibility rather than the coherent flux. The visibility is normalised by the zero-spacing flux, and this allows unknown values of the flux throughput of the system to be factored out. It also allows the use of calibrator stars whose diameters are well known but whose intrinsic brightness at the wavelength of observation are not well known.

In contrast, at mid-infrared wavelengths normalisation of the visibility is made difficult by the large thermal background levels, which make the zero-spacing flux less easy to determine than the coherent flux. As a result, a calibration approach is often adopted, which is similar to that used in radio interferometry: a set of calibrator objects is defined whose coherent fluxes as a function of baseline are well determined, and the calibration is done in terms of coherent fluxes.

For the purposes of this discussion, the convention used at shorter wavelengths, that of using the visibilities, will be used. The principles can be carried over straightforwardly to using coherent fluxes.

The averaged squared visibility can be written in terms of the object visibility and a multiplicative error $\left\langle \left| \gamma_{ij} \right|^2 \right\rangle$ known as the power-spectrum *transfer function* so that

$$\left\langle \left| \hat{V}_{ij} \right|^2 \right\rangle = \left\langle \left| \gamma_{ij} \right|^2 \right\rangle \left| V(\boldsymbol{u}_{ij}) \right|^2. \tag{8.33}$$

The square root of the power-spectrum transfer function is known as the *system visibility*. The simplest assumption is that the system visibility is constant. The transfer function can then be estimated by observing another object (the

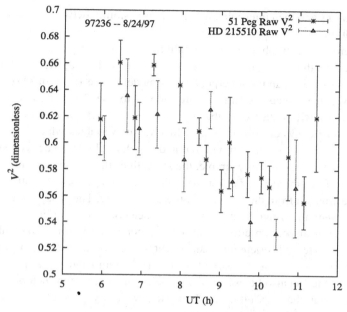

Figure 8.9 Squared visibility measurements from the PTI as a function of time of night. Measurements from both the target star (51 Peg) and the calibrator star (HD 215510) are shown. The authors of the paper conclude that there is little evidence for the target star being resolved, and so the majority of the visibility variation seen in both the target and the calibrator is due to variations in the system visibility. From Boden *et al.* (1998).

calibrator star) whose object visibility function $V(\boldsymbol{u}_{ij})$ is known, typically a star of known diameter. However, the assumption that the system visibility is constant is usually not a good one. An example set of visibility modulus data is shown in Figure 8.9; it can be seen that there may be large variations of the system visibility between different observations of the same star. This is because the system visibility is the product of a number of factors, each of which can vary with time.

The most obvious factor which can change with time is the seeing. Not only are the wavefront perturbations variable from exposure to exposure, but the mean parameters of the seeing r_0 and t_0 also vary with time and with pointing direction. Baldwin *et al.* (2008) report variations of r_0 by a factor of two in the space of less than a second, but typically significant variations in seeing take place on timescales of minutes. Slow variations in seeing can be compensated for by 'bracketing' the observation of the science target with two calibrator observations and then interpolating the system visibility between the calibrator observations.

If there are many observations of calibrator stars during a given observing session, then this technique can be extended to model the system visibility as a function of time and to interpolate this model to the times of observation of the science targets. A further extension is to use the calibrator visibilities together with data from other sources in the interferometer to construct a model of the system visibility as a function of variables which may affect it. Any variable that can be monitored and which might correlate with system visibility can be used. For example, the strength of the temporal seeing can be estimated from the fringe-tracker data while the strength of the spatial seeing can be estimated from the adaptive-optics system data. The group-delay difference between the fringe tracker and the science combiner will likely correlate with spectral dispersion effects and the telescope elevation and azimuth will likely correlate with orientation-dependent telescope vibrations and polarisation effects. Environmental variables such as the temperature and pressure in the delay line can also be used in the model. In many interferometers, all variables that may have an effect on the system visibility are routinely recorded together with the interferometric data.

A multi-dimensional model of the system visibility as a function of these variables can be fitted to the calibrator data, and this model can then be used to estimate the system visibility during the observations of the science targets. The more calibrator observations that are available, the more variables can be used in the model, and so this technique has been used with most success on interferometers that are able to make large numbers of such observations a night, for example the Mark III interferometer on Mt Wilson and the PTI (Colavita *et al.*, 1999).

The error in calibration of the system visibility is often the largest source of measurement error for the visibility modulus, and so estimating this error is critical to later stages of data reduction. Large calibration datasets allow the calibration error to be determined from the misfit between the system visibility model and the data it is based on. With careful modelling these errors can be reduced to around 1% (Mozurkewich *et al.*, 2003).

8.9 OIFITS files

The data-reduction process for an interferometer will have many features which are instrument-specific, but the product at the end of this process has been standardised between interferometers, in the form of a file format.

This format is called OIFITS and is documented by Pauls *et al.* (2005). The format allows for the storage of calibrated power spectra and closure phases and also for the storage of calibrated averaged visibilities – these can be derived from coherent averaging or used to represent spectral differential phase data.

9

Model fitting and image reconstruction

The OIFITS files produced at the end of the data-reduction process can be used to provide information about the object under study. There are two ways in which this information can be extracted, either in terms of parameters of a relatively simple model, or as a model-independent image. These two forms of information are related to one another and are often used in tandem. This chapter discusses the process of deriving these end products of an interferometric observation.

9.1 Bayesian inference

In Chapter 8 the overall inverse problem of interferometry was presented as the problem of determining the parameters of a model of the object being observed from the measured data values. The data-reduction process presented in that chapter does not fundamentally change that problem: the object model parameters still need to be determined, but, as the name implies, the data-reduction process reduces the volume of data that need to be considered.

Indeed, the remaining problem is not one of data reduction but of data interpretation, as the number of data points produced by the data-reduction process may be comparable to or even less than the number of model parameters. This kind of under-constrained problem is where a form of inference known as Bayesian inference is at its best.

Bayesian inference acknowledges the non-uniqueness of the model parameters in the majority of inverse problems. Instead of selecting a single model (for example a parameterised model together with a single set of values for the model parameters) which could have produced the data, it acknowledges that there may be many possible models and assigns a *probability P* between

0 and 1 to all of them in parallel. The probability assigned to a given model is used to denote the plausibility of the model based on all the available evidence, with 0 denoting impossible and 1 denoting certain. Bayesian probability extends Boolean logic, which can only deal with statements that are either true or false, to allow statements about which models are more likely to be true and which are more likely to be false.

Bayes' theorem is the fundamental 'engine' of Bayesian inference. It combines statements about previous knowledge with statements about experimental data to allow new statements to be made about the best models to explain the data. Bayes' theorem states that, given a set of data D, the probability $P(M|D)$ (the *posterior probability* or *posterior* for short) of a particular model M is given by

$$P(M|D) = \frac{P(M)P(D|M)}{P(D)}.$$
(9.1)

This theorem expresses the logic behind the scientific method in a mathematical formalism. It says that the plausibility of any given model after performing an experiment is a combination of the plausibility of the model before the experiment was performed $P(M)$ (the *prior probability* or *prior*) with an experimental update $P(D|M)$, known as the *likelihood* of the data, which expresses how well the the data fits with the predictions of the model. The denominator $P(D)$ is a normalising factor, which ensures that the posterior probability sums to unity; in more advanced Bayesian inference it is called the 'Bayesian evidence' and can be used to distinguish between different classes of model, but it is not considered further here.

A full treatment of Bayesian inference in data analysis can be derived from a number of excellent introductions to the topic, for example that by Sivia and Skilling (2006). The following sections consider the likelihood and the prior probability in the context of interferometry, and show how these can be used together with Bayes' theorem for both model-fitting and imaging.

9.2 The interferometric likelihood

The likelihood $P(D|M)$ expresses how often a given model M will produce a given set of data D. In a perfect experiment, a single model can only produce a single possible set of measurements so $P(D|M)$ would be either unity if the data agree with the model or zero if the data disagree. In this case, all models which could not produce the observed data automatically de-select themselves via Bayes' theorem, as the posterior probability for such models would be zero.

In practice, all realistic models must include the effects of measurement noise, and so the agreement between the data and a given model is only probabilistic – a given model could possibly produce many different sets of measured values, some of these sets being more likely than others.

Determining the likelihood requires a process for producing the data from the model, in other words the forward problem discussed in Section 8.2. The forward problem introduced there starts with the model of the object with a set of parameters $\{\Theta_m\}$ and ends with the individual pixel intensity measurements in the fringe pattern $\{r_p\}$. The forward problem considered in this chapter is an abbreviated one as shown in Figure 9.1. This problem takes as its end point the OIFITS file produced by the data-reduction process described in Chapter 8. In other words, the data reduction moves the 'data' of the forward problem 'backwards' from a set of pixel intensities to an OIFITS file containing a set of calibrated averaged power-spectrum and bispectrum data, and/or a set of calibrated coherently averaged visibilities. The revised forward problem starts from the object model to produce object coherent fluxes $F(u_{ij})$ as before. In order to cover the general case where the data may contain coherently integrated visibilities and not just bispectrum and power-spectrum data, the measurement model incorporates a set of antenna-based phase distortions $\{\epsilon_1, \ldots, \epsilon_M\}$. These phase distortions are so-called 'nuisance parameters': parameters of the forward problem whose value we are not interested in but must be included in the measurement model in order to explain the data. In the full forward problem the set of nuisance parameters includes a complete description of the atmospheric wavefront distortions for every exposure; in this 'data-reduced' model there is only one set of piston phases for each coherently averaged data point, so the number of such nuisance parameters is greatly reduced.

The phase distortions combine with the object coherent fluxes to yield measured coherent fluxes

$$F_{ij} = e^{i[\epsilon_i - \epsilon_j]} F(u_{ij}). \tag{9.2}$$

Note that the system visibility in this model is unity as any system-visibility variations are assumed to have been removed in the calibration process. The data are processed to produce power-spectrum and bispectrum data. The phase errors have no effect on these data, and so do not need to be included in the model if coherently averaged data are not present in the OIFITS file. Finally, the noiseless data are corrupted by additive Gaussian noise to produce the noisy but calibrated data in the OIFITS file. Importantly, it is assumed that

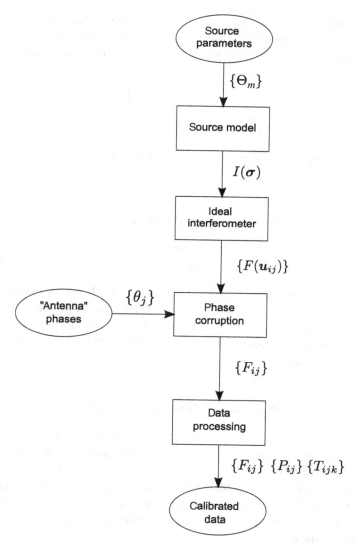

Figure 9.1 The abbreviated forward problem for model fitting.

the noise is uncorrelated between data samples, which is not always the case, as discussed in Section 9.6.4.

The likelihood of the data for a given model can be straightforwardly derived from the noiseless data if the noise is assumed to be Gaussian and uncorrelated between the data points. If the noiseless data, whether bispectrum, power-spectrum and/or coherently averaged visibility data are denoted by by the set $\{\mu_1, \ldots, \mu_N\}$, then the likelihood of a given set of measured data $D = \{d_1, \ldots, d_N\}$ will be given by

$$P(D|M) \propto \prod_{i=1}^{N} e^{-\frac{|d_i - \mu_i|^2}{2\sigma_i^2}}, \qquad (9.3)$$

where σ_i is the standard deviation of the noise added to μ_i. This can be written as

$$P(D|M) \propto e^{-\chi^2}, \qquad (9.4)$$

where χ^2 is the weighted sum of squared deviations:

$$\chi^2 = \sum_{i=1}^{N} \frac{|d_i - \mu_i|^2}{2\sigma_i^2}. \qquad (9.5)$$

Thus, the maximum of the likelihood occurs for models which minimise the 'distance' between the noiseless data predicted by the model and the actual data, i.e. at the minimum of χ^2. Choosing a model which minimises χ^2 is equivalent to least-squares fitting of models to data, and is a form of *maximum likelihood* estimation.

9.3 Model-dependent priors

Bayesian inference extends the ideas of maximum-likelihood estimation by considering the probabilities of the whole model space, not just the single point where the probability is maximised, and also by considering the prior probability as well as the likelihood. The prior depends, as might be expected from the name, on what prior information exists at the time of making the measurements. A specific example serves to illustrate the general process of constructing the prior.

In this example the object under study is a binary star system that is too small to be resolved by conventional telescopes. One possible model for this system is a physical model which has an orbit with a semi-major axis, eccentricity, inclination and so forth, and has masses, ages and metalicities of the constituent stars. This model would be appropriate for fitting to the data from observations at multiple epochs over the orbit of the binary, complemented with spectroscopic and photometric data.

When analysing data from a single epoch on its own, a model with the smallest number of parameters needed to model the expected brightness distribution on the sky might be easier to work with. One such model might consist of a pair of stars, modelled as uniform disks with fluxes at the wavelength of observation of I_A and I_B and angular diameters θ_A and θ_B, respectively. The angular separation θ_{AB} of the pair of stars and the orientation ϕ_{AB} of this separation vector on the sky completes the set of model parameters.

Often the prior probability can be factorised into a set of independent probabilities

$$P(I_A, I_B, \theta_A, \theta_B, \theta_{AB}, \phi_{AB}) = P(I_A)P(I_B)P(\theta_A)P(\theta_B)P(\theta_{AB})P(\phi_{AB}) \qquad (9.6)$$

but this is not always the case. For instance, if the total flux from the system $I_{\text{total}} = I_A + I_B$ is known a priori from single-telescope measurements then I_A and I_B are not independent variables. Instead, the prior probability distribution concentrates along a line in the (I_A, I_B) plane given by $I_A = I_{\text{total}} - I_B$.

Assuming that the total flux is not known and the rest of the prior is also factorisable, the problem of determining the model prior reduces to one of determining priors for each of the variables independently. The role of the prior is easiest to understand when it is used to exclude regions of model space which are unfeasible. For example, based on centuries of research we know that it is impossible (or at least extremely unlikely) for a star to emit negative flux so models with negative I_A or I_B would have zero prior probability.

Assigning values to the prior outside of the impossible regions of parameter space is less straightforward. A simple solution occurs in the case of maximum lack of knowledge: if pre-existing knowledge provides no reason to prefer any one from a number of different models, then all these models are assigned equal probabilities – the so-called *uniform prior*. In the absence of any information about the orientation ϕ_{AB} of the separation vector of the binary pair on the sky, any orientation is equally likely. Thus, a uniform prior over the range $-\pi$ to π can safely be chosen. For the same reason, uniform priors are good choices for the nuisance parameters ϵ_i in the measurement model.

It may be possible to constrain the separation of the binary pair θ_{AB} to be in some given range $\theta_0 \pm \sigma$ if there is some other information available, such as an approximate distance to the system and an orbital period from spectroscopy. In this case an appropriate prior for θ_{AB} might be a Gaussian with a mean of θ_0 and a standard deviation of σ.

If there is no observational evidence to favour any particular scale for θ_{AB} versus a scale which is ten times smaller or ten times larger than this then an appropriate prior would be one which is uniform in the logarithm of θ_{AB}. In other words, the prior $P(\theta_{AB}) \propto 1/\theta_{AB}$ would be appropriate (this is known in Bayesian inference as the 'uninformed prior for a scale parameter').

To avoid infinities in this distribution it is necessary to choose some upper and lower bounds for θ_{AB} outside which the probability tends rapidly towards zero. The upper bound could be the angular separation at which the binary pair would be resolvable with a single telescope while the lower bound could be the angular separation at which the stellar surfaces would be touching for any reasonable choice of system parameters.

Similar log-uniform priors could be used for the fluxes and angular diameters as these are also *scale parameters* – a key attribute these share is that they always take a value which is greater than zero. The exact choice of the cutoffs in such priors is usually unimportant, since if there are a reasonable amount of data then the likelihood will usually overpower these choices in determining the posterior.

9.4 Model fitting in practice

In principle, Bayesian solution of the inverse problem consists simply of computing the posterior probability using Equation (9.1) for every point in the model space. The resulting distribution can be summarised in terms of the location and extent of the region of highest probability. To exhaustively cover all of the multi-dimensional model space would take near-infinite amounts of computer time, but there exist computational techniques for minimising the time spent in regions of model space with low probability and interpolating the probability in higher-probability regions. The details of these techniques can be ignored for the purposes of this discussion and the problem can be treated as though we were in possession of an infinitely large computer, which can compute the probabilities for all points in parallel.

Model-fitting software is often purpose-written for a given astrophysical model, so that all of the physics of the problem can be incorporated. There exist also more general-purpose programs, for example LITpro (Tallon-Bosc *et al.*, 2008) and MFIT (see the supplementary online material referenced in Appendix B), which fit geometrical models. These models can consist of unresolved point sources, disks, shells and so forth.

Model-fitting programs read in calibrated data from an OIFITS file and typically perform either a least-squares fit to the data or find the location of the maximum posterior probability. The output therefore typically consists of a single-point estimate of the model parameters, and some indication of the region in model space around which χ^2 is within some bounds.

Using such programs is relatively straightforward providing only simple models are used. The main difficulty that can arise is if the posterior probability has multiple local maxima, which can cause problems with many optimisation programs, as they can become 'trapped' in these minima. To diagnose this situation it is often useful to look at plots of the actual data against the noiseless data predicted by the model, to see if there are obvious regions of systematic misfit. Discarding the data in these regions or downweighting these and restarting the model fit can give insight into whether the data or the model is at fault. Another technique is to start the model fit from a number of different starting models to see if the solution converges to the same answer every time.

9.5 Model-independent priors

A Bayesian inference framework requires a model for the object being observed. If the physics of the object being studied is well understood, this model can be a relatively simple model with few parameters. Such models have the advantage that relatively little data are needed to adequately constrain the model parameters. Having fitted this model to the data, the best-fit model can be displayed as an image, hence the term *model-dependent imaging*.

The opposite end of the spectrum occurs when relatively little is known a priori about the object under study, such that a whole host of quite different physical and/or geometric models are consistent with existing knowledge. One approach in this situation would be to construct a set of parameterised models for all of these possibilities and then see which of these models the interferometric data are consistent with. Providing an exhaustive set of such models may prove infeasible and so an alternative approach is to try to reconstruct an image which is independent of any specific model of the object, just as would be possible in conventional imaging using a camera.

In Bayesian terms this can be expressed as a 'model' which consists of a number of pixels with unknown brightnesses. This model has the advantage of generality, but the downside is that the number of model parameters is then as large as the number of pixels, which could easily be in the millions. This poses two problems. There is a computational problem of dealing with models of this large dimensionality, but methods of dealing with this are relatively well understood. A more serious problem is that the model is likely to be underconstrained by the data. The number of independent coherent flux measurements in a typical optical interferometric experiment is typically less than 1000 or so, and so cannot possibly constrain a million-pixel image on its own.

Additional constraints are therefore needed in order to 'regularise' the image reconstruction process. In Bayesian image reconstruction, these constraints are embedded in the prior, and this form of regularisation will be assumed in the following discussion. Even given this regularisation, the data need to be not only numerous but sufficiently diverse to avoid degeneracies in the image, that is to say to avoid the situation where many significantly different images fit the data. The following sections look at the interaction of the constraints from the data with the constraints from the a priori information in a quantitative (but not rigorous) fashion, before discussing how model-independent image reconstruction is achieved in practice.

9.6 Preconditions for imaging

9.6.1 Sampling the (u,v) plane

In Section 2.3 it was shown that reconstruction of an $N \times N$ pixel image was possible in principle if the object visibility function is sampled on an $N \times N$ square grid in the (u, v) plane. In practice, visibility samples are never taken on a regular grid, as even if the telescopes are arranged on the ground in a regular grid, baseline projection effects would distort the grid differently for different objects and for different times of night.

Most interferometers have only a few telescopes and so can only sample the visibility function at a few (u, v) locations at a time. Earth rotation can be used to 'sweep out' ellipses on the (u, v) plane, and, if enough time is available, different array configurations can be used on different nights. Even so, the resulting coverage will still be quite uneven as shown in Figure 9.2.

The question then becomes whether, given some quasi-random sampling in the (u, v) plane, there is adequate information to reconstruct a reliable image of the target. One way to address the problem is via the idea of 'invisible distributions'.

An interferometer samples the object visibility function at a set of discrete locations $\{u_i\}$ in the (u, v) plane. At all other locations the visibility function is unknown. Assuming there are no atmospheric phase errors and no other sources of noise affecting the measurements, the coherent flux $F(u)$ corresponding to the true brightness distribution $I(\sigma)$ will clearly fit the measured data exactly. However, there is an infinite set of other object brightness distributions, which will also fit the data exactly. These will be brightness distributions that are the sum of the true object brightness distribution and any distribution $I_{\text{invis}}(\sigma)$ which has the property that

$$F_{\text{invis}}(u_i) = \mathcal{F} \{I_{\text{invis}}(\sigma)\} (u_i) = 0 \qquad (9.7)$$

for all the (u, v) points at which the data have been measured, but is nonzero elsewhere. The brightness distribution I_{invis} is known as an invisible

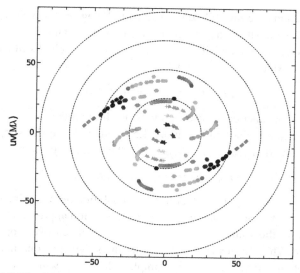

Figure 9.2 The (u, v) plane coverage of observations of VY CMa with the VLTI and the PIONIER instrument. From Monnier *et al.* (2014).

distribution because an object with this brightness distribution would be invisible to an interferometer with the (u, v) coverage given by $\{u_i\}$.

The existence this degeneracy in the image-reconstruction process may make the possibilities for image reconstruction look hopeless. However, there exist a-priori constraints on the invisible distributions, which can be used to 'regularise' the problem. The first of these is image positivity. It is known that no object can emit light with a negative intensity and so all plausible images must be non-negative everywhere. A large fraction of the set of invisible distributions will have negative regions, and if these correspond to regions where the true target is zero, this will lead to an unphysical image.

A second constraint is the finite angular extent of the target (also known as *finite support*). Most astronomical targets can be assumed to consist of a comparatively bright target surrounded by a black background. The angular extent of the illuminated region of the target (the 'support' of the target) can be estimated for many objects from other information about the object. For example, the temperature of an object in thermal equilibrium with the radiation from a star will fall with distance from the star, and so at infrared wavelengths this can be used to estimate the maximum extent of thermal emission from objects in the star's environment. In addition, images of the object taken with single telescopes serve to provide an upper limit to the angular extent.

This constrains the invisible distributions to being ones which do not have any light emission outside a given region centred on the target. This in turn can be used to constrain the invisible distributions. An invisible distribution with a finite angular extent θ_d can be modelled as

$$I_{\text{invis}}(\boldsymbol{\sigma}) = g(\boldsymbol{\sigma})\text{rect}(|\boldsymbol{\sigma}|/\theta_d), \qquad (9.8)$$

where $g(\boldsymbol{\sigma})$ is a function whose extent is not limited and $\text{rect}(x)$ is the unit 'top-hat' function defined in Equation (1.58), which limits the support of the invisible distribution. The coherent flux will therefore be given by

$$F_{\text{invis}}(\boldsymbol{u}) = G(\boldsymbol{u}) * \text{jinc}(\pi|\boldsymbol{u}|\theta_d), \qquad (9.9)$$

where $G(\boldsymbol{u}) = \mathcal{F}\{g(\boldsymbol{\sigma})\}$. The effect of the convolution with the jinc function is to 'smooth' the Fourier-plane function G over a region approximately $1.22/\theta_d$ in size, and so the values of $F_{\text{invis}}(\boldsymbol{u})$ will be correlated over approximately this distance. The Fourier transform of the invisible functions must be zero at the sample locations $\{\boldsymbol{u}_i\}$, and this implies that F_{invis} will have a restricted magnitude over a set of circular regions of order $1/\theta_d$ in radius around the sample points.

If the sampling is such that these regions around the sample points cover most of the (u, v) plane out to some maximum spatial frequency then the invisible distribution will have 'nowhere to hide'. Simulations confirm this general rule that images can be reconstructed from a relatively arbitrary sampling, provided that the samples are spaced at roughly $1/\theta_d$ intervals. In addition, these simulations show that metrics for the quality of image reconstructions from a given (u, v) sampling are typically related to the size of the largest 'hole' in sampling, i. e. the size of the largest region where the Fourier transforms of the invisible distributions are unconstrained.

9.6.2 Phase information

The image reconstruction results quoted above were assuming a phase-stable interferometer, i. e. ones for which both the visibility modulus and phase can be measured. In many early optical interferometers, beams from only two telescopes could be combined at any one time, so only the visibility modulus and not the closure phase could be measured. The question arose whether with sufficient modulus information images could be reconstructed. The answer seems to be that so-called 'phaseless imaging' is possible with high-quality data and/or simple scenes to image, but image reconstruction from phaseless data with scenes of unknown complexity and the levels of systematic errors, random errors and missing data encountered in real life has proven elusive.

Figure 9.3 Upper row: images of two founders of interferometry – H. Fizeau (a) and A. A. Michelson (b). Lower row: hybrid images made from combining the Fourier amplitudes of Fizeau and the Fourier phases of Michelson (c) and from combining the Fourier phases of Fizeau and the Fourier amplitudes of Michelson (d).

This indicates that the phase contains information which is vital for imaging. A simple illustration of this can be derived with the aid of a numerical experiment. In this experiment we take two images such as the upper two images in Figure 9.3 and take their Fourier transforms. If the two image intensities are $i_1(x, y)$ and $i_2(x, y)$, respectively, we can denote the Fourier transforms as

$$A_1(u, v)e^{i\phi_1(u,v)} = \mathcal{F}\{i_1(x, y)\}$$

and

$$A_2(u, v)e^{i\phi_2(u,v)} = \mathcal{F}\{i_2(x, y)\}.$$

We can then make a hybrid of the two images by combining the Fourier phases from one image with the corresponding Fourier amplitudes from the second and taking the inverse Fourier transform:

$$i_{12}(x, y) = \mathcal{F}^{-1}\left\{A_1(u, v)e^{i\phi_2(u,v)}\right\}.$$

We can also make the complementary pairing i_{21}, which has the amplitudes from i_2 and the phases from i_1. The surprising result can be seen in Figure 9.3, namely that the hybrid images strongly resemble the image from which their phase information is derived and bear no resemblance to the image from which

Table 9.1 *The fraction of the phase information captured by the closure
phases in an array of M telescopes; this is the ratio of the number of
independent closure phases N_{clos} to the number of independent object phases
$N_\phi = M(M-1)/2$ measured by a phase-stable interferometer with the same
number of telescopes. The number of independent closure phases is given by
$N_\phi - N_\epsilon$ where $N_\epsilon = M - 1$ is the number of independent antenna-based
errors.*

M	N_ϕ	N_ϵ	N_{closure}	$N_{\text{closure}}/N_\phi$
2	1	1	0	0.00
3	3	2	1	0.33
4	6	3	3	0.50
5	10	4	6	0.60
10	45	9	36	0.80
15	105	14	91	0.87
20	190	19	171	0.90

their amplitude information is derived. This indicates that most of the useful
information in an image is contained in its Fourier phases and not the Fourier
amplitudes: the amplitudes give information about how *much* structure is in an
image, but the phases tell us *where* that structure is.

9.6.3 Information loss in the closure phase

While the closure phase is dependent only on the object visibility phase and
not on the atmospheric phase disturbance, it is not immediately clear how to
make use of the object phase information contained in the closure phase. It is
clear that because of the 'antenna-based' phase errors, some information has
been lost compared to measuring the phases directly.

The amount of information loss can be estimated by comparing the number
of independent closure phase measurements available from an array of M telescopes to the number of phase measurements from the equivalent phase-stable
interferometer. The larger the number M of telescopes in the array, the more
efficient the closure phase is at capturing object phase information, because the
number of object phases rises as $M(M-1)/2$ while the number of atmospheric
phase unknowns rises as $N_\epsilon = M - 1$. It can be seen from Table 9.1 that the
fraction of the object phase information which is retained in the closure phases
tends rapidly towards unity as the number of telescopes increases.

One piece of information which cannot be recovered from closure phases
is information about the absolute position of an object. This can be shown

by considering independent observations of an arbitrary object brightness distribution $I(\sigma)$ and the same distribution displaced by some vector offset σ_1, given by

$$I'(\sigma) = I(\sigma - \sigma_1) = I(\sigma) * \delta(\sigma - \sigma_1), \tag{9.10}$$

where the $*$ operator denotes convolution. From the convolution theorem for Fourier transforms, the corresponding object visibility functions $V(u)$ and $V'(u)$ are related through

$$V'(u) = V(u)e^{-2\pi i\sigma_1 u}. \tag{9.11}$$

As a result, the phases of the visibility functions will be related to one another via

$$\phi'(u) = \phi(u) - 2\pi i\sigma_1 u. \tag{9.12}$$

The measured closure phases on a triangle of telescopes denoted by p, q and r are therefore related by

$$\begin{aligned}
\Phi'_{pqr} &= \phi_{pq} - 2\pi\sigma_1 \cdot u_{pq} + \phi_{qr} - 2\pi\sigma_1 \cdot u_{qr} + \phi_{rp} - 2\pi\sigma_1 \cdot u_{rp} \\
&= \Phi_{pqr} - 2\pi\sigma_1(u_{pq} + u_{qr} + u_{rp}),
\end{aligned} \tag{9.13}$$

where u_{pq} is the spatial frequency corresponding to the baseline between telescopes p and q. For a closed triangle of baselines, $u_{pq} + u_{qr} + u_{rp} = 0$ and so $\Phi_{pqr} = \Phi'_{pqr}$. Thus, all the closure phases measured on a given object are the same as those measured on the same object displaced, and so the absolute position of the object cannot be measured using the closure phase alone.

9.6.4 Using all the closure phases

For an array of M telescopes, there are $\frac{1}{6}M(M - 1)(M - 2)$ different baseline triangles, but the number of linearly independent *object* closure phases (i.e the number of constraints on the phase of the object Fourier transform) that can be measured is only $\frac{1}{2}(M - 1)(M - 2)$. In other words, in a noise-free instrument there would be no point in computing all the possible closure phases since they could all be derived from a linearly independent subset. In the presence of noise, however, it would be worthwhile computing the full set of closure phases if the errors on the different closure phases were uncorrelated, since we would then gain in the number of independent constraints on the object closure phases.

The correlation of two closure phases is the ratio of their covariance to the geometric mean of their variances. The covariance of the closure phases can be calculated from the covariance of the noise on the bispectrum values resolved along lines in the complex plane, which are perpendicular to their respective mean vectors. This calculation can be broken down into the calculations of two covariances:

$$\text{covar}_1[A, B] \equiv E(AB^*) - E(A)E(B^*) \tag{9.14}$$

and

$$\text{covar}_2[A, B] \equiv E(AB) - E(A)E(B), \tag{9.15}$$

where $E(X)$ denotes taking the expectation (mean) of the quantity X. The covariance along any two directions can then be calculated from

$$\text{covar}[A, B, \theta_A, \theta_B] = \frac{1}{2}\text{Re}\{\text{covar}_1[A, B]e^{-i(\theta_A - \theta_B)}$$
$$+ \text{covar}_2[A, B]e^{-i(\theta_A + \theta_B)}\}, \tag{9.16}$$

where $\text{covar}[A, B, \theta_A, \theta_B]$ is the covariance of two complex variables A and B along lines in the complex plane at angles of θ_A and θ_B to the real axis.

The calculations proceed along the same lines as that for computing the variance of the bispectrum in Section 5.5.3. The covariance is zero if the closure triangles do not share a common baseline, but if they do share a common baseline the closure phase covariance is given by

$$\frac{\text{covar}[T_{123}, T_{234}, \phi_{123} + \pi/2, \phi_{234} + \pi/2]}{|F_{12}F_{23}^2 F_{31}F_{34}F_{42}|} = \gamma_{23}^{-2},$$

where γ_{23} is the SNR of the coherent flux on the shared baseline. The correlation coefficient is therefore

$$\mu_{123,234} \equiv \frac{\text{covar}[T_{123}, T_{234}, \phi_{123} + \pi/2, \phi_{234} + \pi/2]}{(\text{var}[T_{123}, \phi_{123} + \pi/2]\text{var}[T_{234}, \phi_{234} + \pi/2])^{1/2}}$$
$$= \frac{1}{3 + 3\gamma^{-2} + \gamma^{-4}},$$

where it has been assumed for simplicity that all the signal-to-noise ratios (SNRs) are the same, i.e. $\gamma_{ij} = \gamma$. Hence,

$$\mu_{123,234} \simeq \begin{cases} 1/3 & \text{at high SNRs} \\ \gamma^4 \simeq 0 & \text{at low SNRs} \end{cases}.$$

Thus, for data where the SNR of the single-exposure coherent flux is high, the noise on the closure phases is strongly correlated and therefore the full set of closure phases contains no new information compared to a linearly independent subset. When the SNR on each exposure is low, however, the noise on the closure phases becomes decorrelated and so using the full set of closure phases will provide a larger number of independent constraints on the object phase.

The above result holds for the average of many bispectrum measurements: for any one exposure, the closure-phase errors cannot all be independent from one another as they arise from the same set of phases. The resolution of this apparent contradiction is that a lack of *correlation* only implies *independence* when the variables involved have Gaussian distributions, and the noise on the bispectrum values in a single exposure do not have Gaussian distributions. The central limit theorem means that the average of many per-exposure bispectrum data results in Gaussian noise and therefore independent data.

9.6.5 Phase reconstruction from the closure phase

If the closure phase can be used to reconstruct the individual object visibility phases, then the equivalence between phase information and closure-phase information is clear. This can be done if the configuration of collectors in the interferometer is appropriately chosen. One such configuration is the linear chain of telescopes such as shown in the lower half of Figure 2.5. If the projected spacing between adjacent telescopes is given by $u_0\lambda$ then the closure phase measured on the linear 'triangle' between telescopes 1, 2, and 3 will be given by

$$\Phi_{123} = \phi(u_0) + \phi(u_0) - \phi(2u_0), \tag{9.17}$$

where $\phi(u)$ is the phase of the object visibility function at spatial frequency u. Thus, if $\phi(u_0)$ is set to some arbitrary value, for example zero (this is equivalent to choosing an arbitrary position for the object, which as shown above cannot be determined from the closure phases), then the value of $\phi(2u_0)$ can be solved for using

$$\phi(2u_0) = 2\phi(u_0) - \Phi_{123}. \tag{9.18}$$

Similarly, the value of closure phase on the next-larger triangle is given by

$$\Phi_{134} = \phi(u_0) + \phi(2u_0) - \phi(3u_0) \tag{9.19}$$

and so the value of $\phi(3u_0)$ can be solved for, giving

$$\phi(3u_0) = 3\phi(u_0) - \Phi_{123} - \Phi_{134}. \tag{9.20}$$

For an array consisting of a longer chain of telescopes, this process can be repeated recursively to determine the phases $\phi(4u_0)$, $\phi(5u_0)$ and so on, up to the phase for the maximum baseline measured by the array. These phases can then be combined with the visibility moduli measured on the corresponding baselines to form a set of estimates of the visibility equivalent to those measured by a phase-stable interferometer.

This procedure, known as *redundant-spacing calibration* shows that closure-phase measurements can act as an effective substitute for phase measurements. However, it does require a special array configuration: as the name suggests, arrays which allow this kind of phase reconstruction must have a substantial degree of baseline redundancy; that is, there are repeated measurements of the visibility of the same spatial frequency by different pairs of telescopes. This means that the number of distinct spatial frequencies measured by the interferometer is less than that measured by a non-redundant interferometer where all telescope pairs measure a unique spatial frequency.

9.7 Model-independent image reconstruction

Given sufficient (u, v) coverage and phase information, it would seem tempting to reconstruct images by continuing with the 'estimator' style of data reduction, reconstructing the image from the calibrated Fourier data through a series of steps. Such processes have been tried, using the closure phase to reconstruct a set of phases and combining these with the visibilities estimated from the power spectrum, inverse Fourier transforming to get an image and then 'cleaning' the image to impose constraints such as non-negativity and finite extent. These processes have met with some success, but have generally fallen out of favour with respect to more Bayesian image reconstruction methods, which directly fit the image to the noisy data.

Bayesian methods are more robust to missing data, for example not requiring redundant telescope arrays in order to allow phase reconstruction and allowing the use of Fourier coverage with significant 'holes'. There are a number of such Bayesian methods (or Bayesian-style methods) for image reconstruction in optical interferometry. Those in active use at the time of writing include (in alphabetical order) programs such as BSMEM (Buscher, 1993), IRBis (Hofmann *et al.*, 2014), MACIM (Ireland *et al.*, 2006), MIRA (Thiébaut, 2008), SQUEEZE (Baron *et al.*, 2010) and WISARD (Meimon *et al.*, 2005), but many more are under development and the state of the art is constantly improving.

9.7.1 BSMEM

The BSMEM program is described here, not to advocate it as the best solution but to provide an example to illustrate the process of image reconstruction. The BiSpectrum Maximum Entropy Method (BSMEM) program was designed, as its name suggests, to reconstruct images from bispectrum (as well as power-spectrum data).

The image reconstruction process can be understood in terms of a standard Bayesian model-fitting where the model consists of a number of image pixels on a regular square grid. The model parameters are the brightnesses of these pixels and in a typical image reconstruction there will be thousands if not millions of such parameters. The number of calibrated data points from an optical interferometer will measure in the dozens to the hundreds at most, and so the problem is highly underconstrained.

In BSMEM this problem is 'regularised', that is additional constraints are introduced in order to make the problem well-posed, using the principle of maximum entropy. This principle will be introduced in this discussion in terms of a Bayesian prior; in reality, despite the fact that the maximum entropy principle was introduced in a Bayesian context, the exact link between maximimum entropy methods and Bayesian analysis remains controversial. More sophisticated analyses suggest that maximum-entropy methods can be better thought of as having Bayesian analysis as a special case, rather than vice versa (Skilling, 1988a; Giffin, 2008).

These conceptual issues are glossed over here and instead a pseudo-Bayesian rationale for the image reconstruction method adopted is presented, which is plausible if not looked at too closely. The reader is asked to suspend disbelief on the grounds that more rigorous rationales require considerably more space to explain; the interested reader can consult the literature (Skilling, 1988b; Gull and Skilling, 1999). The proof of the maximum-entropy pudding is that it has been successfully used in many different fields to give results superior in many ways to other existing methods.

The maximum-entropy prior for a set of pixels with intensities $\{i_1, \ldots, i_{N_{pix}}\}$ is given by

$$P(I, \alpha | M) \propto e^{\alpha S}, \tag{9.21}$$

where S is the relative 'entropy' of the image with respect to a prior model image $M = \{m_1, \ldots, m_{N_{pix}}\}$ and given by

$$S = -\sum_{p=1}^{N_{pix}} \left[i_p - m_p - i_p \log(i_p/m_p) \right]. \tag{9.22}$$

The parameter α is a 'hyperparameter' of the prior and controls how strongly the entropy affects the posterior probabilities. The model image M encodes prior information about the image, for example the finite support constraint can be encoded as low values of m_p outside some region, indicating that little or no flux resides in these 'blank spaces'.

Given the maximum-entropy prior, Bayes' theorem gives the posterior probability as

$$P(I, \alpha | M, D) \propto e^{\alpha S - \chi^2}. \tag{9.23}$$

A convenient representation for the multi-dimensional posterior is the single image I, which has the maximum posterior probability, in other words the image which maximises $\alpha S - \chi^2$.

For a given value of α the probability can be maximised by maximising the entropy S and by minimising the misfit with the data χ^2. The former term comes from the prior and the latter comes from the likelihood. The relative importance of these two terms is given by the hyperparameter α. If α is small, the maximisation tends towards the minimum of the misfit χ^2 between the actual data and the noiseless data predicted from the image I. If α is large, the maximisation tends towards the maximum of the entropy S, which occurs when the image is identical to the prior image M. Using too small a value of α can over-fit the data at the expense of the prior while using too large a value of α tends to emphasise the prior at the expense of the data.

A balance between these two can be achieved by choosing the value for α such that the maximal value of $\alpha S - \chi^2$ occurs when $\chi^2 = N_{dat}$. This is equivalent to maximising the entropy under the constraint that $\chi^2 = N_{dat}$. This value of χ^2 occurs when the distance between the actual data and those which would be observed if I were the true image is of order of one standard deviation per data point.

If the model M is uniform over some region and zero elsewhere, the net effect of this choice for α is that the selected image I is the 'smoothest' image that fits the data, as the entropy penalises images which have more structure. In any interferometric experiment (and indeed in any imaging experiment), the measured Fourier components do not extend beyond some finite limit in the (u, v) plane, and so the object can have arbitrary amounts of small-scale structure without affecting the data. The maximum-entropy method automatically limits any attempts to 'guess' at this structure while at the same time providing an image that shows any structure for which there is good evidence.

The maximum-entropy prior also has the useful property of enforcing positivity in the image – the entropy contributed by any pixel tends towards $-\infty$ as the pixel intensity tends towards zero and so pixel values are 'steered' away from negative values by the entropy.

9.7.2 Running BSMEM

The BSMEM program implements the entropy-maximisation described above for data D, which consist of an arbitrary mixture of bispectrum and power-spectrum data – at present it cannot make use of coherently-integrated visibility data directly. The data, together with the errors on the data, are read in from an OIFITS file specified by the user.

The user also needs to specify a model image M. The default model is a broad Gaussian centred in the middle of the image, which can be used to signify that the object is thought to have an extent that is smaller than some characteristic contour (such as the half-maximum contour) of the Gaussian but where there is not a definite cut-off in the maximum size of the possible structure. This encourages the algorithm to find images which fit inside this contour but does not completely exclude slightly larger images.

An arbitrary model image can also be loaded from a user-provided file. This is typically derived from lower-resolution data – sometimes it is derived from a previous image reconstruction phase using the same data, which the user has adjusted and smoothed to indicate plausible areas for the flux in the object to reside. This allows the user to strengthen the limit-support constraint based on prior knowledge of what structures are likely to exist in the object under study, and allows images to be reconstructed from more sparsely sampled data than looser constraints. Needless to say, such user adjustment requires some care to avoid introducing spurious structure into the image.

The program works in terms of an $N \times N$-pixel image I where each pixel represents a square region on the sky of a fixed angular size. This pixel size can be automatically selected by the program or chosen by the user. Typically, pixels which are a few times smaller than the theoretical maximum resolution of the data given by u_{max}^{-1} are used. In other words, the pixels are smaller than the 'resels' (resolution elements) of the instrument. This allows for some 'super-resolution', where details can be recovered in the image which are below the conventional resolution limit if there is sufficient evidence in the data for such features. The 'smoothing' effect of the maximum-entropy prior will ensure that if no such super-resolution is possible then no spurious sub-resel structure will be put in. The size of the image in terms of the number of pixels N needs to be set large enough to include the maximum possible extent of the image based on prior information.

Given these initial parameters, the BSMEM program uses iterative gradient-descent algorithms to find the constrained maximum-entropy solution, starting from the model image M. The progress of this gradient descent in terms of the values of χ^2 and the entropy can be monitored during the running of the program and intermediate images can be plotted. A situation often arises that the algorithm stagnates because it cannot find an image which fits the data adequately. This is most likely because such an image does not exist, because the noise on the data has been underestimated. In this case it may help to edit the data to put in more realistic error estimates, or to remove data points which appear (for justifiable reasons) to be outliers.

9.7.3 Image reconstruction from coherently averaged data

Data which have been coherently averaged can straightforwardly be converted into bispectrum and power-spectrum data for use in image reconstruction programs which expect these as input. However, the noise model assumed for the data will be wrong since, for example, the bispectrum amplitudes and the power-spectrum amplitudes will be strongly correlated, and the closure phases will be correlated as explained in Section 9.6.4.

There are many image-reconstruction programs that can be used directly with coherently averaged visibility data. These are mostly written with data from radio interferometers in mind, but the forward model for coherently averaged data in an optical interferometer is the same as the forward model for a phase-unstable radio interferometer and so the noise model used is appropriate. Image reconstruction from data corrupted by antenna-dependent phase errors is known in radio interferometry as *self-calibration* (Pearson and Readhead, 1984), referring to the fact that external calibration for these phase errors is not needed.

The program VLBMEM (Sivia, 1987) illustrates the general technique, but other image-reconstruction programs exist that are based on similar principle. The VLBMEM program uses an iterative 'hybrid mapping' scheme to solve for two sets of model parameters: the image pixels $\{i_p\}$ and the antenna-dependent phase errors $\{\epsilon_i\}$. There are two steps in each iteration, one attempting to solve for each set. In the first step, the model image from the previous iteration (or a point-source starting image) is used together with the visibility data to solve for the phase errors $\{\epsilon_i\}$. This is achieved through a least-squares fit between the noiseless visibilities predicted from the model image $\left\{V(\boldsymbol{u}_{ij})e^{i[\epsilon_i-\epsilon_j]}\right\}$ and the measured data $\left\{V_{ij}\right\}$, with the set of $\{\epsilon_i\}$ as free parameters. In the second step, the $\{\epsilon_i\}$ are held fixed at the values output from the previous step and the image pixels $\{i_p\}$ are adjusted to fit the data using a maximum-entropy method. These two steps are repeated until convergence is achieved at a self-consistent solution for the image and the corrupting antenna-based phases.

Other self-calibration programs exist for dealing with phase-corrupted radio interferometric data. They typically use the same two-step process but differ in the way to solve for the image pixels in the second stage. A popular method for this 'deconvolution' step is the CLEAN algorithm (Hogbom, 1974).

9.8 Image quality

The images reconstructed from interferometric data can be of high quality when the (u, v) sampling is adequate and the SNR of the visibility data is high. Figure 9.4 shows a series of infrared (2-μm wavelength) images of the dust shell around the carbon star IRC+10216. These images were reconstructed from visibility-modulus and closure-phase data from aperture-masking observations made at the Keck telescope (Tuthill *et al.*, 2000a).

The images show complex structures and most of these structures are repeated in all the images, despite being taken at different times, with different aperture masks, different filters and different rotations of the masks. Epoch-to-epoch differences seen in these images can in many cases be attributed to dust motion. Thus, the basic reliability of these image reconstructions is not in doubt, despite the fact that several possible sources of ambiguity, including incomplete (u, v) coverage and the use of closure phases rather than visibility phases, are present.

Nevertheless, like all experimental data, these images are not perfect representations of what the object looks like. First, the image resolution is limited by the baseline coverage. The maximum baseline used in these reconstructions is about 10 m, (limited by the size of the Keck telescope) and so the nominal resolution is $\lambda/B_{max} \approx 40$ milliarcseconds. The images show structures on scales perhaps slightly smaller than this, indicating some level of super-resolution in the image reconstruction.

Secondly, the structure at lowest contour levels, at the level of 1% of the peak brightness, is perhaps less reliable, showing epoch-to-epoch variations which are unlikely to be physical changes in the source but rather are likely to be noise in the image. Thus the 'dynamic range' of these images, which is defined as the range of the peak brightness in the image to the brightness of the weakest believable features, is perhaps between 50:1 and 100:1.

The dynamic range is the most commonly used metric of the 'fidelity' of an interferometric image, that is to say how well the image approximates to the true object brightness distribution. It is a useful metric for understanding how suitable the reconstructed images are for answering scientific questions such as the detectability of faint companions.

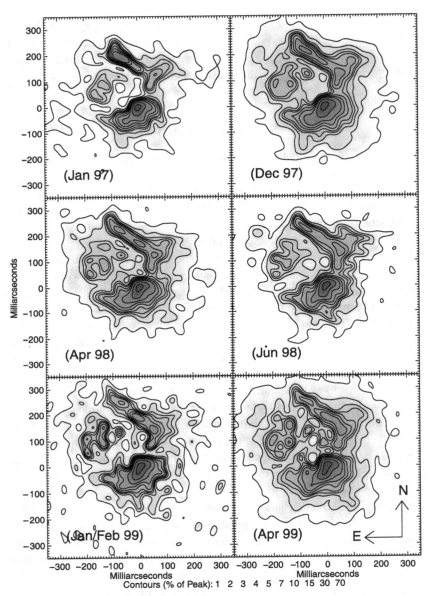

Figure 9.4 Images at several epochs of the evolved star IRC+10216, showing the complex structure of the dust shells around the star. From Tuthill *et al.* (2000a).

The dynamic range has the advantage of being readily applicable to experimental data where the true object brightness distribution is not available for comparison, but where there are regions of blank sky within the imaged region which are unlikely to contain any flux. Any features seen in this region can be taken to be the noise level and the ratio of the peak flux in the image to the flux in these features will give a measure of the dynamic range. However, some caution must be taken in interpreting the dynamic range derived in this way as the effect of regularisation such as that produced by maximum-entropy methods can suppress the level of noise peaks.

A rule of thumb for the dynamic range that can be expected from an interferometric image can be obtained from a simple model of the imaging process. In this model, the data $\{F_{ij}\}$ are taken at N_{uv} sample points in the (u, v) plane using a phase-stable interferometer. The data are corrupted by zero-mean additive noise $\{n_{ij}\}$ so that

$$F_{ij} = F(\boldsymbol{u}_{ij}) + n_{ij}. \tag{9.24}$$

The noisy synthesised image, in other words the inverse Fourier transform of the noisy data, will be given by

$$I_{\text{noisy}}(\boldsymbol{\sigma}) = \hat{I}(\boldsymbol{\sigma}) + \sum_{i,j} \left(n_{ij} e^{2\pi i \boldsymbol{u}_{ij} \cdot \boldsymbol{\sigma}} + n_{ij}^* e^{-2\pi i \boldsymbol{u}_{ij} \cdot \boldsymbol{\sigma}} \right), \tag{9.25}$$

where $\hat{I}(\boldsymbol{\sigma})$ is the noiseless synthesised image given in Equation (2.21). The two noise terms in Equation (9.25) come from the fact that every coherent flux sample at location $\boldsymbol{u}_{ij}$ also gives a measurement of the value at $-\boldsymbol{u}_{ij}$ due to the Hermitian symmetry of the Fourier transform of a real function.

If the coherent flux noise distribution has circular symmetry in the complex plane and the noise values are uncorrelated, then the noise at any location in the synthesised image is twice the real part of a 'random walk' in the complex plane. The standard deviation of the noise at a given point in the image will

$$\sigma_{\text{image}} = \sqrt{2 \sum_{i,j} \sigma_{ij}^2}, \tag{9.26}$$

where the factor of $\sqrt{2}$ arises from taking twice the real part of a noise distribution, which is circularly symmetric.

If the object is a point source so that the coherent flux modulus $|F(\boldsymbol{u})| = F$ everywhere in the (u, v) plane, then the peak flux in the synthesised image will be given by

$$I_{\text{max}} \approx 2N_{\text{uv}} F \tag{9.27}$$

while if the noise is constant so that $\sigma_{ij} = \sigma$,

$$\sigma_{\text{image}} = \sqrt{2N_{\text{uv}}}\sigma. \tag{9.28}$$

Thus, the dynamic range of the image will be given by

$$\text{dynamic range} \sim \frac{2N_{\text{uv}}F}{\sqrt{2N_{\text{uv}}}\sigma} = \sqrt{2N_{\text{uv}}}\text{SNR}_F, \tag{9.29}$$

where

$$\text{SNR}_F = \frac{F}{\sigma}. \tag{9.30}$$

Thus, the dynamic range is proportional to the SNR of the coherent flux data and the square root of the number of data points.

This rule of thumb has been found to give a good indication of the dynamic ranges of images reconstructed from closure-phase and visibility-modulus data, providing that the (u, v) sampling is adequate. For the images in Figure 9.4, the number of (u, v) samples going into the reconstruction is of order 100, so it can be inferred that the average SNR of the coherent flux samples is probably of order 5. Given that this object is very bright the SNR inferred is likely to be representative of the typical random calibration error level rather than being due to detection noise.

If the (u, v) sampling is not adequate for the field-of-view and/or the noise is highly correlated, for example because of systematic calibration errors, then the dynamic range computed from Equation (9.29) may not give a good estimate of the reliability of the image. Figure 9.5 shows a set of ten image reconstructions from the same interferometric dataset, taken on the dust-enshrouded supergiant VY CMa (Monnier *et al.*, 2014). The data have approximately 2% errors on the visibility modulus and 1° errors on the closure phases, corresponding to an SNR of order 50. About 300 different (u, v) samples were taken. Using Equation (9.29) would yield a dynamic-range estimate of order $\sqrt{2 \times 300 \times 50} \approx 170 : 1$, implying a high-quality image.

In reality it can be seen that on this object different reconstruction packages (or even the same reconstruction package used by different teams) tend to agree on the strongest small-scale features but disagree to quite a large extent on the larger-scale features. This is likely due to a lack of short-baseline data: the object was previously imaged using aperture masking and this showed structure out to scales of order 200 milliarcseconds, but the data used in these reconstructions have a shortest baseline such that $|u| > 5 \times 10^6$ (the (u, v) coverage is shown in Figure 9.2), and this corresponds to a spatial scale of order 40 milliarcseconds. As a result, the large-scale flux is poorly constrained and so the image reconstruction packages have a difficult time determining it.

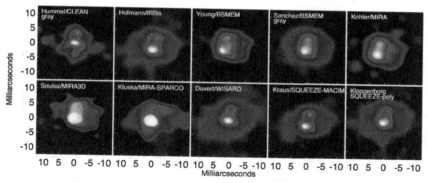

Figure 9.5 Image reconstructions of the dust around the supergiant VY CMa from the 2014 image 'beauty contest' (Monnier *et al.*, 2014).

Thus, even with apparently high-quality data, the reliability of the image reconstruction can be seriously compromised if the (u, v) sampling is not appropriate for the object. This must be borne in mind whenever interpreting any interferometric image reconstruction.

9.9 Practical case study: imaging and model-fitting on Betelgeuse

When analysing observational data, imaging and model-fitting are not mutually exclusive activities but are complementary. Crude model-fitting can be used to guide image reconstruction while the results from image reconstruction can be used to decide on which models to fit. This can be illustrated by using a practical example of imaging of Betelgeuse using aperture-masking data. The data presented are from 1990, but illustrate the scientific progress that can be made using relatively simple images when they have an angular resolution which is much higher than has been previously possible.

Betelgeuse is the nearest supergiant star, and as a result has one of the largest apparent angular diameters of any star in the sky. It was the first star to have its angular diameter measured using interferometry (Michelson and Pease, 1921). Supergiant stars are quite unlike main-sequence stars such as the Sun in that not only are they thousands of times larger, but the size of the convection cells at the surface of the star is comparable to the size of the star itself. Thus, the stellar surface might be expected to be pocked, not with the diminutive dark sunspots we see on the Sun, which are caused by localised cooling due to the magnetic field, but with giant bright 'starspots' corresponding to convective

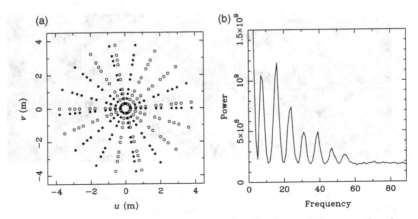

Figure 9.6 (a) The Fourier plane coverage of the Betelgeuse 710-nm wavelength observations. (b) The average power spectrum of 500 interferograms taken on Betelgeuse at 546-nm wavelength. These data were taken using a five-hole mask, but fringes are only visible on the shortest seven out of ten baselines because the source is resolved on longer baselines.

upwelling of hotter material from below the surface of the star. Despite this intrinsic interest, no convincing image of the features on the surface of this star had been made at the time of observation.

The observations were made using aperture masking on the 4.2-m William Herschel Telescope. The longest baseline in the aperture mask was 3.8 m and so at a wavelength of 700 nm the highest spatial frequency measured corresponded to a sinusoid on the sky with a peak-to-peak spacing of 38 mas. Betelgeuse has an angular diameter of 40–50 mas so the stellar surface is only just resolved.

The mask was arranged to be non-redundant so no two pairs of holes had the same spacing: for a five-hole mask fringes appeared at ten different frequencies. To achieve two-dimensional Fourier plane coverage the whole arrangement was rotated around the telescope axis to a number of different position angles with respect to the sky. With about ten different position angles and ten baselines per position angle, Fourier data at about 100 (u, v) points were obtained as shown in Figure 9.6. These are not all independent visibility points, but the resulting image has less than 10×10 independent resolution elements in it, so the coverage is adequate.

The observations at each wavelength yielded of order 100 calibrated visibility amplitudes and 100 closure phases. Figure 9.7 shows a subset of these data for illustration purposes. A number of important constraints on the resulting image can be inferred purely from examination of these data (a kind of 'eyeball model fitting'):

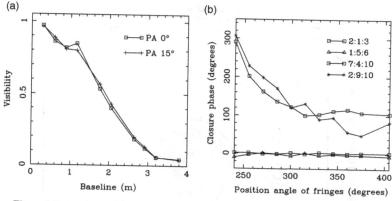

Figure 9.7 A subset of the calibrated visibility (a) and closure-phase (b) measurements on Betelgeuse at a wavelength of 710 nm. The visibility data are plotted for two similar position angles of the mask. The closure phase data are plotted for two sets of triples containing only short baselines and two sets of triples containing long baselines. The closure phases are labelled with the indices (1–10) of the baselines making up the triple.

- From the decrease in calibrated visibility with baseline it is clear that Betelgeuse is indeed resolved on the longest baselines and this is roughly consistent with a uniform disc diameter of around 50 milliarcseconds.
- The closure phases measured on baseline triangles involving the longest baselines show large departures from 0° and 180°, whereas the closure phases on the triangles containing only short baselines are close to zero. This means that there is significant asymmetric structure on scales comparable with the size of the stellar disc – the asymmetry cannot be due to a companion far away from the star.
- The closure phase varies over essentially 0° to 360°, indicating that the asymmetric structure is comparable to or larger than the symmetric structure at this resolution. The fraction of the flux that is unresolved at this resolution is given approximately by the visibilities on these baselines: these visibilities are of order 10%, hence the flux in the asymmetric structure is of the order of 10% of the total flux.

An additional inference which can be drawn is that the closure phase can be measured with very few systematics: the uncalibrated closure phases measured on the shortest baselines where Betelgeuse is effectively unresolved are within a degree of zero. This kind of accuracy in a conventional phase measurement would require an internal pathlength stability of better than 2 nm and yet no special precautions were taken in this experiment.

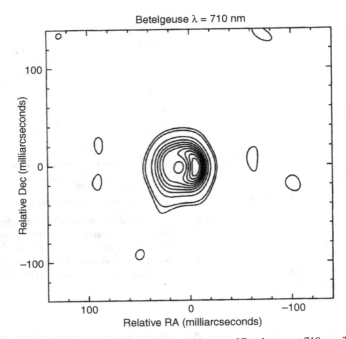

Figure 9.8 Contour map of the reconstructed image of Betelgeuse at 710 nm. The contours are at 1, 2, 10, 20, 30, ... , 90% of the peak intensity. From Buscher *et al.* (1990).

Image reconstruction on these data was performed with the VLBMEM program described in Section 9.7.3. This program was designed for radio inter-ferometry data, and so expects coherently averaged data, but only incoherently averaged power-spectrum and bispectrum data were available.

The procedure used to adapt the incoherently-averaged data to the program was one which is a common way of using optical interferometry data with radio-interferometry software. This was to create a set of 'fake' coherently averaged data that were consistent with the bispectrum and power-spectrum data. This used a least-squares fit of a set of visibility phases to the closure phases. Note that the VLBMEM program assumes that the phases are cor-rupted by unknown antenna phases, so it was not necessary to solve for the antenna phases at this stage. The visibility modulus was derived from the square root of the calibrated power-spectrum estimates.

This procedure is less preferable than using a program which is designed to accept bispectrum and power-spectrum data directly, since the noise model used is not as accurate. Nevertheless, it produced image reconstructions of Betelgeuse (Figure 9.8) which agree with the constraints inferred from the

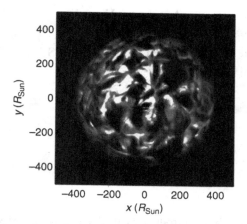

Figure 9.9 Image of a red supergiant star simulated using three-dimensional radiative hydrodynamic modelling. Note the size of the star – the axes are labelled in units of the Sun's radius. From Chiavassa *et al.* (2011).

data. The reconstructed image shows a stellar disc of about 50 milliarcseconds diameter and a region of increased intensity on one side of the disc.

Having made a model-independent image reconstruction, one can then fit simple models to the data to get quantitative information. These model-fits show that the 'hotspot' seen on the disc does indeed account for about 10% of the stellar flux. Further aperture-masking observations (Wilson *et al.*, 1997) showed that such hotspots appear and disappear on timescales of order several weeks and observations in the infrared (Young *et al.*, 2000) showed that these spots have significantly lower contrast at longer wavelengths.

The combination of image reconstruction and the model-fitting can then be used to constrain astrophysical models. Radiative hydrodynamic modelling of supergiants has shown that turbulent upwelling processes in stars like Betelgeuse can produce bright hotspots with the characteristics seen in these interferometric observations, and produces simulated images such as that shown in Figure 9.9. These images can be compared to the hotspots seen in interferometric images in different wavebands to constrain the models for the star.

Further understanding of these remarkable stars will require detailed images of these turbulent structures, preferably with sufficient time resolution to follow their evolution. Interferometry provides the only way to make 'movies' of structures like this, and hopefully future interferometers will provide us with pictures as beautiful and intriguing as seen in these simulations.

Appendix A

Fourier transforms

The following is a brief introduction to the Fourier transform at the level required for this book. More complete descriptions can be found in textbooks, for example in Gaskill (1978) and Bracewell (2000) or online, for example in Wikipedia.

A.1 Fourier series

We start by recalling the mathematics of Fourier series. Consider a periodic function of time $f(t)$, with a period T corresponding to a frequency $v_0 = 1/T$. The function can be written as a superposition of sinusoidal waves with frequencies $v_0, 2v_0, 3v_0, \ldots$:

$$f(t) = \frac{1}{2}A_0 + \sum_{n=1}^{\infty} \left(A_n \cos\left(\frac{2\pi nt}{T}\right) + B_n \sin\left(\frac{2\pi nt}{T}\right) \right). \tag{A.1}$$

This decomposition into sine and cosine waves is known as a Fourier series representation of a function. The set of coefficients A_n and B_n define an alternative repesentation of $f(t)$, since $f(t)$ can be determined uniquely from these coefficients.

To find the value of a particular coefficient A_n or B_n, we multiply the above equation by $\cos\left(\frac{2\pi nt}{T}\right)$ or $\sin\left(\frac{2\pi nt}{T}\right)$ and integrate from $-T/2$ to $T/2$ to give

$$A_n = \frac{2}{T} \int_{-T/2}^{T/2} f(t) \cos\left(\frac{2\pi nt}{T}\right) dt, \tag{A.2}$$

$$B_n = \frac{2}{T} \int_{-T/2}^{T/2} f(t) \sin\left(\frac{2\pi nt}{T}\right) dt. \tag{A.3}$$

These results rely on the fact that

$$\int_{-T/2}^{T/2} \cos\left(\tfrac{2\pi mt}{T}\right) \cos\left(\tfrac{2\pi nt}{T}\right) dt = 0 \qquad \text{for } m \neq n$$
$$= T/2 \quad \text{for } m = n$$

$$\int_{-T/2}^{T/2} \cos\left(\tfrac{2\pi mt}{T}\right) \sin\left(\tfrac{2\pi nt}{T}\right) dt = 0 \qquad \text{for all } m, n$$

$$\int_{-T/2}^{T/2} \sin\left(\tfrac{2\pi mt}{T}\right) \sin\left(\tfrac{2\pi nt}{T}\right) dt = 0 \qquad \text{for } m \neq n$$
$$= T/2 \quad \text{for } m = n.$$

A.1.1 Complex coefficients

Rather than write the series as a sum of sine and cosine terms, it often convenient to describe the Fourier components with complex coefficients which represent the amplitude and phase of each component:

$$f(t) = \sum_{n=-\infty}^{\infty} C_n e^{i2\pi nt/T} = \sum_{n=-\infty}^{\infty} C_n e^{in\nu_0 t}. \qquad (A.4)$$

The coefficients C_n are given by multiplying by $e^{-2\pi i m \nu_0 t}$ and integrating from $-T/2$ to $T/2$:

$$C_n = \frac{1}{T} \int_{-T/2}^{T/2} f(t) e^{-2\pi i n \nu_0 t} dt. \qquad (A.5)$$

This results relies on the fact that

$$\int_{-T/2}^{T/2} e^{2\pi i n \nu_0 t} e^{-2\pi i m \nu_0 t} dt = 0 \qquad \text{for } m \neq n$$
$$= T \quad \text{for } m = n. \qquad (A.6)$$

We are mostly interested in functions $f(t)$ which are real, and in this case we find that $C_{-m} = C_m^*$. This ensures that the imaginary parts of the terms in Equation (A.4) cancel to zero when summing over positive and negative n.

A.2 Generalisation to non-periodic functions

We now need to extend the idea of Fourier series to functions that do not repeat; in other words, aribitrary functions (although there exist functions which cannot be Fourier transformed; these are of more interest to mathematicians than to physicists). Conceptually, this can be seen as taking the limit of a Fourier series as $T \to \infty$. The fundamental frequency ν_0 tends to zero, and so the frequency components making up the series become infinitesimally close together. The discrete coefficients $\{C_m\}$ merge to become a continuous *spectrum* $g(v)$ and the sum becomes an integral, giving

$$f(t) = \int_{-\infty}^{\infty} g(v)e^{2\pi i v t} df, \tag{A.7}$$

where

$$g(v) = \int_{-\infty}^{\infty} f(t)e^{-2\pi i v t} dt. \tag{A.8}$$

These equations define the Fourier transform; $g(v)$ is the Fourier transform of $f(t)$ and $f(t)$ is the inverse Fourier transform of $g(v)$. The reader should be aware that there are alternative definitions of the Fourier transform, which use $\omega = 2\pi v$ instead of v as the 'conjugate variable' to t, but the definitions above (sometimes called the 'unitary' form of the Fourier transform) are used in this book.

We denote the Fourier transform using the operator $\mathcal{F}$, which transforms the function f into the function g

$$\mathcal{F}[f(t)] = g(v), \tag{A.9}$$

and its inverse $\mathcal{F}^{-1}$, which derives f given g:

$$\mathcal{F}^{-1}[g(v)] = f(t). \tag{A.10}$$

The Fourier transform can be thought of in terms of analysis and synthesis: the forward transform splits a function into sinusoidal components at different frequencies, while the inverse transform synthesises a function from these components.

A.3 Generalisation to higher dimensions

Fourier transforms are not just applied to functions of time: they can equally be applied to functions of any variable. For example, we can Fourier transform a function of a spatial variable x to derive a function of spatial frequency s, i.e.

$$\mathcal{F}[f(x)] = g(s). \tag{A.11}$$

The spaces spanned by the variables v and s are 'reciprocal spaces' complementary to t and x, respectively, since they have units of s^{-1} and m^{-1}, respectively.

In the case of spatial dimensions, it is natural to consider generalising the Fourier transform to functions of two- or higher-dimensional variables. This is straightforwardly done by replacing the scalar dimension x by a vector $\mathbf{x}$

and by integrating over multiple dimensions. In two dimensions we have the forward transform,

$$\mathcal{F}[f(x)] = g(s) = \iint_{-\infty}^{\infty} f(x)e^{-2\pi i s \cdot x} \, dx \, dy, \qquad (A.12)$$

where $x = (x, y)$ is a two-dimensional spatial coordinate and $s = (s_x, s_y)$ is a two-dimensional spatial frequency, and the inverse transform,

$$\mathcal{F}^{-1}[g(s)] = f(x) = \iint_{-\infty}^{\infty} g(s)e^{2\pi i s \cdot x} \, ds_x \, ds_y. \qquad (A.13)$$

A.4 Dirac delta functions

An function which is important in the theory of Fourier transforms is the *Dirac delta function*. This function is used to denote a sharp 'spike', which occurs over an infinitesimal time, but with finite area. An example of where this might be used is the idea of an 'impulse' in Newtonian mechanics, where a sharp 'kick' imparts finite momentum even though it occurs over an infinitesimally short time Δt. The momentum change is $F\Delta t$, so as Δt tends to zero, F must tend to infinity during the kick. We would represent the force as a function of time in this limiting case as a Dirac delta function.

Formally, the Dirac delta function $\delta(t)$ is defined by the property

$$\int_{-\infty}^{\infty} \delta(t)f(t) \, dt = f(0) \qquad (A.14)$$

for any arbitrary function $f(t)$. This means that $\delta(t)$ is a unit-area spike at $t = 0$ and is zero everywhere else. The function $\delta(t - t_0)$ is the same spike offset from the origin by an amount t_0 so

$$\int_{-\infty}^{\infty} \delta(t - t_0)f(t) \, dt = f(t_0). \qquad (A.15)$$

In other words, multiplying a function by $\delta(t - t_0)$ and integrating returns a 'sample' of the value of the function at time t_0.

We can use this property to show that the Fourier transform of a delta function is a complex exponential:

$$\int_{-\infty}^{\infty} \delta(t - t_0)e^{-2\pi i \nu t} \, dt = e^{-2\pi i \nu t_0}. \qquad (A.16)$$

Note that the Fourier transform of a delta function at the origin is a constant (i.e. independent of ν).

Delta functions can be defined by analogy in higher dimensions. A point source in two dimensions can be defined as $\delta(x, y)$ such that

$$\iint_{-\infty}^{\infty} \delta(x, y) f(x, y) \, dy \, dx = f(0, 0). \tag{A.17}$$

A.5 Convolution

We now introduce the idea of the *convolution* of two functions $f(y)$ and $g(y)$. The convolution is denoted by the operator '$*$' and defined by the equation

$$f(y) * g(y) = \int_{-\infty}^{\infty} f(u) g(y - u) \, du. \tag{A.18}$$

The easiest example of convolution to visualise is when one of the functions is a delta function:

$$f(y) * \delta(y - y_0) = \int_{-\infty}^{\infty} f(u) \delta(y - y_0 - u) \, du = f(y - y_0). \tag{A.19}$$

In other words, the function $f(y)$ is reproduced centred around the delta function rather than around zero. For general functions, we can think of $g(y)$ as being made up of the sum of an infinite number of delta functions with different 'heights':

$$g(y) = \int_{-\infty}^{\infty} g(u) \delta(y - u) \, du. \tag{A.20}$$

Hence, when we take the convolution of f and g, each of the delta functions making up g is replaced by a copy of f. The function g is therefore 'smeared out' by the function f (and vice versa).

Convolution has an important application in optics where imaging systems such as telescopes are concerned. These instruments are imperfect, so a point-source object (a δ-function) is smeared out in the image. The image produced from a delta-function object is described by the resolution function (or *point-spread function*) of the instrument. For a general object, the image produced is the convolution of the object with the resolution function. The operation of trying to remove the effects of the resolution function is known as *deconvolution*.

Convolutions are particularly useful in the context of Fourier transforms. It can be shown that the Fourier transform of the product of two functions is the convolution of the Fourier transforms of the individual functions, i.e. if

$$F(s) = \mathcal{F}[f(x)] \tag{A.21}$$

and

$$G(s) = \mathcal{F}[g(x)], \tag{A.22}$$

then

$$\mathcal{F}[f(x)g(x)] = F(s) * G(s). \tag{A.23}$$

Similarly, the Fourier transform of the convolution of two functions is the product of the Fourier transforms of the individual functions,

$$\mathcal{F}[f(x) * g(x)] = F(s)G(s). \tag{A.24}$$

A.6 Composing Fourier transforms

Evaluating the Fourier transform or inverse Fourier transform of a function by performing the integrals in Equations (A.7) and (A.8) can be time-consuming. An alternative in many cases is to make use of a few 'building-block' Fourier transforms and combine them using known mathematical properties of the Fourier transform. Some commonly used functions and their transforms are given in Table A.1 and the following results can be used to extend and combine the transforms:

Reciprocity: If we know the forward transform of a function, we can obtain the inverse transform of the same function by 'flipping' the result about the $t = 0$ axis: if the Fourier transform of $f(t)$ is $g(v)$, then the inverse Fourier transform of $f(v)$ is $g(-t)$. This follows from the similarity of the integrals defining forward and inverse transforms, and means that all the following results apply when you replace $\mathcal{F}$ with $\mathcal{F}^{-1}$.

Scaling law: If we 'stretch' a function horizontally by an amount a, the corresponding dimension in the transform is compressed by the same factor: $\mathcal{F}[f(t/a)] = |a|g(av)$. This reciprocal scaling is related to Heisenberg's uncertainty principle, since wavefunctions of quantum variables such as position and momentum form a Fourier-transform pair. Note that the vertical dimension of the transformed function is stretched by $|a|$.

Linearity: If a function is the superposition of two other functions, its Fourier transform is the superposition of the respective Fourier transforms: $\mathcal{F}[a_1 f_1(t) + a_2 f_2(t)] = a_1 \mathcal{F}[f_1(t)] + a_2 \mathcal{F}[f_2(t)]$, where a_1 and a_2 are arbitrary constants and f_1 and f_2 are arbitrary functions.

The convolution theorem: $\mathcal{F}[f_1(t)f_2(t)] = \mathcal{F}[f_1(t)] * \mathcal{F}[f_2(t)]$ and $\mathcal{F}[f_1(t) * f_2(t)] = \mathcal{F}[f_1(t)]\mathcal{F}[f_2(t)]$.

Fourier transforms

Table A.1 *Fourier transforms of some useful functions.*

	$f(t)$		$g(v)$
Delta function	$\delta(t - t_0)$	complex exponential	$e^{-2\pi i v t_0}$
One-dimensional top-hat	$\text{rect}(t)$	sinc	$\frac{\sin(\pi v)}{\pi v}$
Circular top-hat	$\text{rect}(|\boldsymbol{x}|)$	jinc	$\frac{J_1(|s|)}{|s|}$
Gaussian	$e^{-t^2/2}$	Gaussian	$\sqrt{\pi} e^{-\pi v^2}$
Comb function	$\sum_{n=-\infty}^{\infty} \delta(t - n)$	Comb function	$\sum_{n=-\infty}^{\infty} \delta(v - n)$

A.7 Symmetry

It is often helpful to know the symmetry properties of Fourier transforms in order to check that we have got the right results. It is straightforward to show from Equation A.8 that the Fourier transform of a function $f(t)$ that is purely real (something which is true of most of the physical variables we will be taking the Fourier transform of) has so-called *Hermitian symmetry*, i.e. that $g(-v) = g(v)^*$. This means that the properties of the function can be determined purely from the positive frequency components of the Fourier transform, so typically we only plot the positive half of the Fourier transform in these cases.

Similarly, we can show that if a function is both real and symmetric (i.e. $f(-t) = f(t)$) then its Fourier transform is both real and symmetric, while if the function is both real and antisymmetric (i.e. $f(-t) = -f(t)$) then its transform is purely imaginary and antisymmetric.

Appendix B

Supplementary online material

In a rapidly changing field such as interferometry, a static item like a book quickly goes out of date. To supplement the material in this book, some material has been placed online and can be accessed at the URL http://www.cambridge.org/9781107042179.

The online material includes links to other online material such as websites for the major interferometry projects and sites providing interferometry software.

It also includes additional tools based on the material presented in the book. Although many of the figures in this book have been taken from the literature, many had to be prepared from scratch or reworked into a format suitable for presentation in this book. The graphs were prepared using the programming language Python, using its numerical libraries Numpy and SciPy, together with the Matplotlib plotting library. The source code for these plotting programs, together with the graphical results, have been made available as part of the online material.

It is hoped that these will form a valuable technical and educational resource in themselves. In some cases, readers may want to know the exact values that were used in a given graph. Instead of taking a ruler to the graph, they can print out the values with a small modification to the relevant plotting program. In other cases, the reader may want to compare or combine the results from different graphs and, as the software has been written in a relatively modular fashion, this should be straightforward. Access to the source code of the programs should also allow the techniques used in the preparation of the data to be studied in detail and employed elsewhere.

Included in this software is a simple interferometer simulation framework. This includes an atmospheric wavefront perturbation generator of the type described in Section 3.5.2, a means of correcting the perturbations with a simple adaptive optics system and a means of measuring the fringe parameters of spatially filtered and unfiltered fringes. This could be extended in many ways to derive new results and I encourage people to experiment with it.

The software has been released with an open-source licence to encourage the use and further development of the code. Also included are some of the diagrams from this book, licenced under a Creative Commons licence. It is hoped that these could prove useful in putting together a course based on the material in this book.

References

Angel, J. R. P., J. M. Hill, P. A. Strittmatter, P. Salinari and G. Weigelt. Interferometry with the large binocular telescope. *Proc. SPIE*, 3350:881–889, 1998.

Armstrong, J. T., D. Mozurkewich, L. J Rickard *et al.* The Navy Prototype Optical Interferometer. *Astrophys. J.*, 496:550–572, 1998.

Baldwin, J. E., M. G. Beckett, R. C. Boysen *et al.* The first images from an optical aperture synthesis array – mapping of Capella with COAST at 2 epochs. *Astron. Astrophys.*, 306:L13–L16, 1996.

Baldwin, J. E., P. J. Warner and C. D. Mackay. The point spread function in lucky imaging and variations in seeing on short timescales. *Astron. Astrophys.*, 480:589–597, 2008.

Baron, F., D. Monnier and B. Kloppenborg. A novel image reconstruction software for optical/infrared interferometry. *Proc. SPIE*, 7734: doi: 10.1117/12.857364, 2010.

Basden, A. H. and D. F Buscher. Improvements for group delay fringe tracking. *MNRAS*, 357:656–668, 2005.

Benisty, M., J.-P. Berger, L. Jocou *et al.* An integrated optics beam combiner for the second generation VLTI instruments. *Astron. Astrophys.*, 498:601–613, 2009.

Berger, D. H., J. D. Monnier, R. Millan-Gabet *et al.* CHARA michigan phase-tracker (CHAMP): a preliminary performance report. *Proc. SPIE.*, 7013:701319. 1–701319.10, 2008.

Bernat, D., A. H. Bouchez, M. Ireland *et al.* A close companion search around l dwarfs using aperture masking interferometry and Palomar laser guide star adaptive optics. *Astrophys. J.*, 715:724, 2010.

Bessell, M. S., F. Castelli and B. Plez. Model atmospheres broad-band colors, bolometric corrections and temperature calibrations for o-m stars. *Astron. Astrophys*, 333:231–250, 1998.

Birks, T. A., J. C. Knight and P. S. Russell. Endlessly single-mode photonic crystal fiber. *Opt. Lett.*, 22:961–963, 1997.

Boden, A. F., G. T. van Belle, M. M. Colavita *et al.* An interferometric search for bright companions to 51 pegasi. *Astrophys. J. Lett.*, 504:L39, 1998.

Bohec S. Le and J. Holder. Optical intensity interferometry with atmospheric Cherenkov telescope arrays. *Astrophys. J.*, 649(1):399, 2006.

Bracewell, R. N. *Fourier Transform and its Applications*. McGraw Hill, Boston, MA, 3rd edition, 2000.

Breckinridge J. B. Measurement of the amplitude of phase excursions in the Earth's atmosphere. *J. Opt. Soc. Am.*, 66:143–144, 1976.

Buscher, D. F. Optimising a ground–based optical interferometer for sensitivity at low light levels. *MNRAS*, 235:1203–1226, 1988a.

Buscher, D. F. Getting the most out of COAST. PhD thesis, Cambridge University, 1988b.

Buscher, D. F. Direct Maximum-Entropy image reconstruction from the bispectrum. In J. G. Robertson and W. J. Tango, editors, *Very High Angular Resolution Imaging (IAU Symposium 158)*, pages 91–93, Sydney, 1993.

Buscher, D. F. A thousand and one nights of seeing measurements on Mt Wilson. *Proc. SPIE.*, 2200:260–271, 1994.

Buscher, D. F., C. A. Haniff, J. E. Baldwin and P. J. Warner. Detection of a bright feature on the surface of Betelgeuse. *MNRAS*, 245:7P–11P, 1990.

Buscher, D. F., J. T. Armstrong, C. A. Hummel *et al.* Interferometric seeing measurements on Mt. Wilson – power spectra and outer scales. *Appl. Opt.*, 34:1081–1096, 1995.

Buscher, D. F., J. S. Young, F. Baron, and C. A. Haniff. Fringe tracking and spatial filtering: phase jumps and dropouts. *Proc. SPIE.*, 7013:10.1117/12.789869, 2008.

Buscher, D. F., M. Creech-Eakman, A. Farris, C. A. Haniff and J. S. Young. The conceptual design of the Magdalena Ridge Observatory Interferometer. *J. Astron. Instrum.*, 02(02):1340001, 2013.

Caves, C. M. Quantum limits on noise in linear amplifiers. *Phys. Rev. D*, 26:1817–1839, 1982.

Ceus, D., L. Delage, L. Grossard *et al.* Contrast and phase closure acquisitions in photon counting regimes using a frequency upconversion interferometer for high angular resolution imaging. *MNRAS*, 430:1529–1537, 2013.

Chiavassa, A., B. Freytag, T. Masseron and B. Plez. Radiative hydrodynamics simulations of red supergiant stars: IV. Gray versus non-gray opacities. *Astron. Astrophys.*, 535:A22, 2011.

Colavita, M. M., J. K. Wallace, B. E. Hines *et al.* The Palomar Testbed Interferometer. *Astrophys. J.*, 510:505–521, 1999.

Colavita, M. M., M. R. Swain, R. L. Akeson, C. D. Koresko and R. J. Hill. Effects of atmospheric water vapor on infrared interferometry. *PASP*, 116(823):876–885, 2004.

Connes, P. and G. Michel. Astronomical Fourier spectrometer. *Appl. Opt.*, 14:2067–2084, 1975.

Coudé du Foresto, V., P. J. Bordé, A. Mérand *et al.* FLUOR fibered beam combiner at the CHARA array. *Proc. SPIE*, 4838:280–285, 2003.

Dainty, J. C. and A. H. Greenaway. Estimation of spatial power spectra in speckle interferometry. *J. Opt. Soc. Am.*, 69:786–790, 1979.

Dali Ali, W., A. Ziad, A. Berdja *et al.* Multi-instrument measurement campaign at paranal in 2007: characterization of the outer scale and the seeing of the surface layer. *Astron. Astrophys.*, 524:A73, 2010.

Davis, J., P. R. Lawson, A. J. Booth, W. J. Tango and E. D. Thorvaldson. Atmospheric path variations for baselines up to 80 m measured with the Sydney University Stellar Interferometer. *MNRAS*, 273:L53–L58, 1995.

Davis, J., W. J. Tango and E. D. Thorvaldson. Dispersion in stellar interferometry: simultaneous optimization for delay tracking and visibility measurements. *Appl. Opt.*, 37:5132–5136, 1998.

Davis, J., W. J. Tango, A. J. Booth *et al.* The Sydney University Stellar Interferometer – I. The instrument. *MNRAS*, 303:773–782, 1999.

Faucherre, M., B. Delabre, P. Dierickx and F. Merkle. Michelson- versus Fizeau-type beam combination: is there a difference? *Proc. SPIE*, 1237:206–217, 1990.

Ferrari, M., G. R. Lemaitre, S. P. Mazzanti *et al.* VLTI pupil transfer: variable curvature mirrors: I. final results and performances and interferometric laboratory optical layout. *Proc. SPIE*, 4006:104–115, 2000.

Finger, G., I. Baker, D. Alvarez *et al.* Evaluation and optimization of NIR HgCdTe avalanche photodiode arrays for adaptive optics and interferometry. *Proc. SPIE*, 84530:84530T–84530T, 2012.

Fisher, M., R. C. Boysen, D. F. Buscher *et al.* Design of the MROI delay line optical path compensator. *Proc. SPIE.*, 7734: doi: 10.1117/12.857168, 2010.

Fried, D. L. Optical resolution through a randomly inhomogeneous medium for very long and very short exposures. *J. Opt. Soc. Am.*, 56:1372–1379, 1966.

Fried, D. L. The nature of atmospheric turbulence effects on imaging and pseudo-imaging systems, and its quantification. In J. Davis and W. J Tango, editors, *High Angular Resolution Stellar Interferometry*, IAUC 50, American Physical Society, College Park, MD, pages 4–1–4–44, 1978.

Gaskill, J. D. *Linear Systems, Fourier Transforms, and Optics.* Wiley, New York, 1978.

Giffin, A. Maximum entropy: the universal method for inference. PhD, State University of New York at Albany, 2008.

Gordon, J. A. and D. F. Buscher. Detection noise bias and variance in the power spectrum and bispectrum in optical interferometry. *Astron. Astrophys.*, 541:A46, May 2012. doi: 10.1051/0004-6361/201117335.

Greco, V. G. Molesini and F. Quercioli. Telescopes of Galileo. *Applied Optics*, 32 (31):6219, November 1993. ISSN 0003-6935. doi: 10.1364/AO.32.006219.

Gull, S. F. and J. Skilling. Quantified maximum entropy: MemSys5 users' manual, 1999.

Hale, D. D. S., M. Bester, W. C. Danchi, W. Fitelson, S. Hoss, E. A. Lipman, J. D. Monnier, P. G. Tuthill, and C. H. Townes. The Berkeley Infrared Spatial Interferometer: A heterodyne stellar interferometer for the mid-infrared. *Astrophys. J.*, 537:998–1012, 2000.

Hanbury-Brown, R. and R. Q. Twiss. Correlation between photons in two coherent beams of light. *Nature*, 177:27–29, 1956.

Hanbury-Brown, R., J. Davis and L. R. Allen. The stellar interferometer at Narrabri Observatory-I: a description of the instrument and the observational procedure. *MNRAS*, 137:375–392, 1967.

Hanbury-Brown, R., J. Davis, D. Herbison-Evans and L. R. Allen. A study of Gamma 2 Velorum with a stellar intensity interferometer. *MNRAS*, 148:103–117, 1970.

Haniff, C. A. and D. F. Buscher. Speckle imaging with partially redundant masks: preliminary results. In J. M. Beckers and F. Merkle, editors, *Proceedings of High Resolution Imaging by Interferometry II*, ESO, Garching bei München, 1992.

Haniff, C. A., C. D. Mackay, D. J. Titterington *et al.* The first images from optical aperture synthesis. *Nature*, 328:694–696, 1987.

Hofmann, K.-H., G. Weigelt and D. Schertl. An image reconstruction method (IRBis) for optical/infrared interferometry. *Astron. Astrophys.*, 565:A48, 2014.

Hogbom, J. Aperture synthesis with a non-regular distribution of interferometer baselines. *Ap. J. Suppl. Ser.*, 15:417–426, 1974.

Horton, A. J., D. F Buscher and C. A Haniff. Diffraction losses in ground-based optical interferometers. *MNRAS*, 327:217–226, 2001.

Hummel, C. A., D. Mozurkewich, N. M. Elias *et al.* Four years of astrometric measurements with the Mark III optical interferometer. *Astron. J.*, 108:326–336, 1994.

Ireland, M. J. and J. D. Monnier. A dispersed heterodyne design for the Planet Formation Imager. *Proc. SPIE*, 9146:914612–914612, 2014.

Ireland, M. J., J. D. Monnier and N. Thureau. Monte-Carlo imaging for optical interferometry. *Proc. SPIE*, 6268:doi: 10.1117/12.670940, 2006.

Jennison, R. C. A phase sensitive interferometer technique for the measurement of the Fourier transforms of spatial brightness distribution of small angular extent. *MNRAS*, 118:276–284, 1958.

Jorgensen, A. M., H. R. Schmitt, J. T. Armstrong *et al.* Coherent integration results from the NPOI. *Proc. SPIE*, 7734:77342Q–77342Q–13, 2010.

Jovanovic, N., P. G. Tuthill, B. Norris *et al.* Starlight demonstration of the dragonfly instrument: an integrated photonic pupil-remapping interferometer for high-contrast imaging. *MNRAS*, 427:806–815, 2012.

Kellerer, A. and A. Tokovinin. Atmospheric coherence times in interferometry: definition and measurement. *Astron. Astrophys.*, 461:775–781, 2007.

Koechlin, L. The i2t interferometer. In F. Merkle, editor, *Proceedings of NOAO-ESO Conference on High Resolution Imaging by Interferometry*, Garching bei München, ESO, 1988.

Kolmogorov, A. N. The local structure of turbulence in incompressible viscous fluid for very large Reynolds numbers. In *Dokl. Akad. Nauk SSSR*, 30:301–305, 1941.

Korff, D. Analysis of a method for obtaining near-diffraction-limited information in the presence of atmospheric turbulence. *J. Opt. Soc. Am.*, 63:971–980, 1973.

Launhardt, R., T. Henning, D. Queloz *et al.* The ESPRI project: narrow-angle astrometry with VLTI-PRIMA. *Proc. IAU*, 3 (Symposium S248):417–420, 2007.

Lawson, P., editor. *Principles of Long-Baseline Stellar Interferometry*. Jet Propulsion Laboratory, Pasadena, CA, 1999.

Le Bouquin, J.-B. and O. Absil. On the sensitivity of closure phases to faint companions in optical long baseline interferometry. *Astron. Astrophys.*, 541:A89, 2012.

Le Bouquin, J.-B., J.-P. Berger, B. Lazareff, PIONIER: a 4-telescope visitor instrument at VLTI. *Astron. Astrophys.*, 535:A67, 2011.

Lévêque, S., B. Koehler and O. Lühe. Longitudinal dispersion compensation for the very large telescope interferometer. *Astrophys. Space Sci.*, 239:305–314, 1996.

Ma, C., E. F. Arias, T. M. Eubanks *et al.* The international celestial reference frame as realized by very long baseline interferometry. *Astron. J.*, 116:516–546, 1998.

Mahajan, V. N. Strehl ratio for primary aberrations in terms of their aberration variance. *J. Opt. Soc. Am.*, 73:860–861, 1983.

Malvimat, V., O. Wucknitz and P. Saha. Intensity interferometry with more than two detectors? *MNRAS*, 437:798–803, 2014.

Mandel, L. Photon degeneracy in light from optical maser and other sources. *J. Opt. Soc. Am.*, 51:797–798, 1961.

Mandel, L., E C G Sudarshan and E Wolf. Theory of photoelectric detection of light fluctuations. *Proc. Phys. Soc.*, 84:435–444, 1964.

Mariotti, J. M. and S. T. Ridgway. Double Fourier spatio-spectral interferometry: combining high spectral and high spatial resolution in the near infrared. *Astron. Astrophys.*, 195:350–363, 1988.

Martin, F., A. Tokovinin, A. Ziad *et al.* First statistical data on wavefront outer scale at La Silla observatory from the GSM instrument. *Astron. Astrophys.*, 336:L49–L52, 1998.

McGlamery, B. L. Computer simulation studies of compensation of turbulence degraded images. *Proc. SPIE*, 74:225–233, 1976.

Meimon, S. C., L. M. Mugnier and G. Le Besnerais. Reconstruction method for weak-phase optical interferometry. *Opt. Lett.*, 30:1809–1811, 2005.

Mérand, A., P. Bordé and V. Coudé du Foresto. A catalog of bright calibrator stars for 200-m baseline near-infrared stellar interferometry. *Astron. Astrophys.*, 433:1155–1162, 2005.

Michelson, A. A. On the application of interference methods to astronomical measurements. *Astrophys. J.*, 51:257–262, 1920.

Michelson, A. A. and F. G. Pease. Measurement of the diameter of Alpha Orionis with the interferometer. *Astrophys. J.*, 53:249–259, 1921.

Millour, F., O. Chesneau, M. Borges Fernandes *et al.* A binary engine fuelling HD 87643's complex circumstellar environment, determined using AMBER/VLTI imaging. *Astron. Astrophys.*, 507:317–326, 2009.

Monnier, J. D., F. Baron, M. Anderson *et al.* Tracking faint fringes with the CHARA-Michigan phasetracker (CHAMP). *Proc. SPIE*, 8445:84451I–1–84451I–9, 2012.

Monnier, J. D. J.-P. Berger, J.-B. Le Bouquin *et al.* The 2014 interferometric imaging beauty contest. *Proc. SPIE*, 9146:91461Q–91461Q–20, 2014.

Mourard, D., J. M. Clausse, A. Marcotto *et al.* VEGA: Visible spEctroGraph and polArimeter for the CHARA array: principle and performance. *Astron. Astrophys.*, 508:1073–1083, 2009.

Mozurkewich, D., J. T Armstrong, R. B Hindsley *et al.* Angular diameters of stars from the Mark III optical interferometer. *Astron. J.*, 126:2502–2520, 2003.

Nightingale N.S., and D. F. Buscher. Interferometric seeing measurements at the La Palma Observatory. *MNRAS*, 251:155, 1991.

Noll, R. J. Zernike polynomials and atmospheric turbulence. *J. Opt. Soc. Am.*, 66:207–211, 1976.

Pauls, T. A., J. S. Young, W. D. Cotton and J. D. Monnier. A data exchange standard for optical (visible/IR) interferometry. *PASP*, 117:1255–1262, 2005.

Pearson T. J. and A. C. S. Readhead. Image formation by self-calibration in radio astronomy. *Ann. Rev. Astron. Astrophys*, 22:97–130, 1984.

Perrin, G. S. Lacour, J. Woillez and E. Thiébaut. High dynamic range imaging by pupil single-mode filtering and remapping. *MNRAS*, 373:747–751, 2006a.

Perrin, G., J. Woillez, O. Lai *et al.* Interferometric coupling of the Keck telescopes with single-mode fibers. *Science*, 311:194–194, 2006b.

Petrov, R. G., F. Millour, S. Lagarde *et al.* VLTI/AMBER differential interferometry of the broad-line region of the quasar 3C 273. *Proc. SPIE*, 8445:doi: 10.1117/12.926595, 2012.

Porro, I. L., W. A. Traub and N. P. Carleton. Effect of telescope alignment on a stellar interferometer. *Appl. Opt.*, 38:6055–6067, 1999.

Prasad, S. Implications of light amplification for astronomical imaging. *J. Opt. Soc. Am. A*, 11:2799–2803, 1994.

Readhead, A. C. S., T. S. Nakajima, T. J. Pearson *et al.* Diffraction-limited imaging with ground-based optical telescopes. *Astron. J.*, 95:1278–1296, 1988.

Richichi, A. and I. Percheron. First results from the ESO VLTI calibrators program. *Astron. Astrophys.*, 434:1201–1209, 2005.

Roddier, F. The effects of atmospheric turbulence in optical astronomy. In E Wolf, editor, *Progress in Optics*, Elsevier, Amsterdam, 1981, volume 19, pages 281–376.

Sandler, D. G., S. Stahl, J. R. P. Angel, M. Lloyd-Hart and D. McCarthy. Adaptive optics for diffraction-limited infrared imaging with 8-m telescopes. *J. Opt. Soc. Am. A*, 11:925–945, 1994.

Schöller, M. The Very Large Telescope Interferometer: current facility and prospects. *New Astron. Rev.*, 51:628–638, 2007.

Ségransan, D., P. Kervella, T. Forveille and D. Queloz. First radius measurements of very low mass stars with the VLTI. *Astron. Astrophys.*, 397:L5–L8, 2003.

Shaklan, S. and F. Roddier. Coupling starlight into single-mode fiber optics. *Appl. Opt.*, 27:2334–2338, 1988.

Shao, M., M. M. Colavita, B. E. Hines *et al.* The Mark III stellar interferometer. *Astron. Astrophys.*, 193:357–371, 1988.

Shao, M. SIM: the space interferometry mission. *Proc. SPIE*, 3350:536–540, 1998.

Simohamed, L. M. and F. Reynaud. A 2 m stroke optical fibre delay line. *Pure Appl. Opt.: J. Eur. Opt. Soc. A*, 6:L37, 1997.

Sivia, D. S. and J. Skilling. *Data Analysis: A Bayesian Tutorial*. Oxford University Press, Oxford, 2nd edition, 2006.

Sivia, D. S. Phase extension methods. PhD thesis, Cambridge University, 1987.

Skilling, J. The axioms of maximum entropy. In *Maximum-Entropy and Bayesian Methods in Science and Engineering*, Springer, Dordrecht, 1988a, pages 173–187.

Skilling, J. Classic maximum entropy. In J. Skilling, editor, *Maximum Entropy and Bayesian Methods*, 45–52. Springer, Dordrecht, 1988b.

Stephan, E. Sur l'extreme petitesse du diametre apparent des etoiles fixes. *C. R. Acad. Sc. (Paris)*, 78:1008–1112, 1874.

Strano, G. Galileo's telescope: history, scientific analysis, and replicated observations. *Exp. Astron.*, 25:17–31, 2009.

Stürmer, J. and A. Quirrenbach. Simulating aperture masking at the Large Binocular Telescope. *Proc. SPIE*, 8445:84452H–1–84452H–8, 2012.

Tallon-Bosc, I., M. Tallon, E. Thiébaut *et al.* LITpro: a model-fitting software for optical interferometry. *Proc. SPIE*, 7013:70131J–70131J, 2008.

Tango, R. J. Dispersion in stellar interferometry. *Appl. Opt.*, 29:516–521, 1990.

Tango, R. J. and R. Q. Twiss. Diffraction effects in long path interferometers. *Appl. Opt.*, 13:1814–1819, 1974.

Tango, R. J. and R. Q. Twiss. Michelson stellar interferometry. *Progr. Opt.*, XVII:239–277, 1980.

Tatarski, V. I. *Wave Propagation in a Turbulent Medium*. McGraw-Hill Book Company, Inc, New York 1961.

Tatulli, E. and G. Duvert. AMBER data reduction. *New Astron. Rev.*, 51:682–696, 2007.

Tatulli, E., F. Millour and A. Chelli, Interferometric data reduction with AMBER/VLTI. Principle, estimators, and illustration. *Astron. Astrophys.*, 464:29–42, 2007.

ten Brummelaar, T. A. Differential path considerations in optical stellar interferometry. *App. Opt.*, 34:2214–2219, 1995.

ten Brummelaar, T. A., H. A. McAlister, S. T. Ridgway *et al.* First results from the CHARA array. II. A description of the instrument. *Astrophys. J.*, 628:453–465, 2005.

Thiébaut, E. MIRA: an effective imaging algorithm for optical interferometry. *Proc. SPIE*, 7013, doi: 10.1117/12.788822, 2008.

Thompson, A. R., Moran, I. M. and G. W. Swenson Jr. *Interferometry and Synthesis in Radio Astronomy*, 2nd edition, John Wiley & Sons, New York, 2008.

Thureau, N. Compensation of longitudinal dispersion for the GI2T-REGAIN optical interferometer. *J. Op. A: Pure Appl. Opt.*, 3:440, 2001.

Thureau, N. D., R. C. Boysen, D. F. Buscher *et al.* Fringe envelope tracking at COAST. *Proc. SPIE.*, 4838:956–963, 2003.

Traub, W. A. Polarization effects in stellar interferometers. In *NOAO-ESO Conference on High-Resolution Imaging by Interferometry: Ground-based Interferometry at Visible and Infrared Wavelenghts*, ESD, Gerching bei München, volume 29, pages 1029–1038, 1988.

Tubbs, R. Effect of wavefront corrugations on fringe motion in an astronomical interferometer with spatial filters. *Appl. Opt.*, 44:6253–6257, 2005.

Tuthill, P. G. The unlikely rise of masking interferometry: leading the way with 19th century technology. *Proc. SPIE.*, 8445:844502-1–844502-11, 2012.

Tuthill, P. G., J. D. Monnier, W. C. Danchi and B. Lopez. Smoke signals from IRC +10216. I. Milliarcsecond proper motions of the dust. *Astrophys. J.*, 543:284, 2000a.

Tuthill, P. G., J. D. Monnier, W. C. Danchi, E. H. Wishnow and C. A. Haniff. Michelson interferometry with the Keck I telescope. *PASP*, 112:555–565, 2000b.

van Cittert, P. H. Die wahrscheinliche Schwingungsverteilung in einer von einer lichtquelle direkt oder mittels einer Linse beleuchteten Ebene. *Physica*, 1:201–210, 1934.

van Dam, M., E. Johansson, P. Stomski *et al.* Performance of the Keck II AO system. Technical Report 489, W. M. Keck Observatory, 2007.

Wagner, R. E. and W. J. Tomlinson. Coupling efficiency of optics in single-mode fiber components. *Appl. Opt.*, 21:2671–2688, 1982.

Wang, J. Y. and J. K. Markey. Modal compensation of atmospheric turbulence phase distortion. *J. Opt. Soc. Am.*, 68:78–87, 1978.

Wheelon, A. D. *Electromagnetic Scintillation. I. Geometrical Optics*. Cambridge University Press, Cambridge, 2001.

Wilson, R. W. and C. R. Jenkins. Adaptive optics for astronomy: theoretical performance and limitations. *MNRAS*, 278:39–61, 1996.

Wilson, R. W., V. S. Dhillon and C. A. Haniff. The changing face of Betelgeuse. *MNRAS*, 291:819+, 1997.

Woan, G. and P. J. Duffett-Smith. Determination of closure phase in noisy conditions. *Astron. Astrophys.*, 198:375, 1988.

Young, J. S., J. E. Baldwin, R. C. Boysen *et al.* New views of Betelgeuse: multi-wavelength surface imaging and implications for models of hotspot generation. *MNRAS*, 315:635–645, 2000.

Zernike, F. The concept of degree of coherence and its application to optical problems. *Physica*, 5:785–795, 1938.

Zhao, M., J. D. Monnier, E. Pedretti *et al.* Imaging and modeling rapidly rotating stars: α Cephei and α Ophiuchi. *Astrophys. J.*, 701:209–224, 2009.

Index

Printed in the United States
by Baker & Taylor Publisher Services